Nietzsche and The Earth

Also Available from Bloomsbury

Nietzsche and Epicurus, ed. Vinod Acharya and Ryan J. Johnson
Nietzsche and Friendship, Willow Verkerk
Understanding Nietzsche, Understanding Modernism, ed. Brian Pines and Douglas Burnham
Nietzsche and The Antichrist, Daniel Conway

Nietzsche and The Earth

Biography, Ecology, Politics

Henk Manschot

Translated by
Liz Waters

BLOOMSBURY ACADEMIC
LONDON • NEW YORK • OXFORD • NEW DELHI • SYDNEY

BLOOMSBURY ACADEMIC
Bloomsbury Publishing Plc
50 Bedford Square, London, WC1B 3DP, UK
1385 Broadway, New York, NY 10018, USA
29 Earlsfort Terrace, Dublin 2, Ireland

BLOOMSBURY, BLOOMSBURY ACADEMIC and the Diana logo
are trademarks of Bloomsbury Publishing Plc

First published in the Netherlands as *Blijf de aarde trouw*
Pleidooi voor een nietzscheaanse terrasofie © 2016 Henk Manschot and Uitgeverij Vantilt, Nijmegen

First published in Great Britain 2021
This paperback edition published in 2022

Copyright © Henk Manschot, 2021

Henk Manschot has asserted his right under the Copyright,
Designs and Patents Act, 1988, to be identified as Author of this work.

Liz Waters has asserted her right under the Copyright,
Designs and Patents Act, 1988, to be identified as Translator of this work.

For legal purposes the Acknowledgements on pp. xv–xviii constitute an
extension of this copyright page.

Cover design by Charlotte Daniels
Cover image: Flora and Fauna of Africa from Meyers Konversations-Lexikon, 1896.
Illustration: Mützel (© Grafissimo / Getty Images)

All rights reserved. No part of this publication may be reproduced or transmitted in any
form or by any means, electronic or mechanical, including photocopying, recording, or
any information storage or retrieval system, without prior permission in writing from the publishers.

Bloomsbury Publishing Plc does not have any control over, or responsibility for, any thirdparty
websites referred to or in this book. All internet addresses given in this book were
correct at the time of going to press. The author and publisher regret any inconvenience
caused if addresses have changed or sites have ceased to exist, but can accept no
responsibility for any such changes.

A catalogue record for this book is available from the British Library.

Library of Congress Cataloging-in-Publication Data

Names: Manschot, Henk, author. | Waters, Liz, translator.
Title: Nietzsche and the earth: biography, ecology, politics / Henk Manscho; translated by Liz Waters.
Other titles: Blijf de aarde trouw. English
Description: London; New York: Bloomsbury Academic, 2020. | "First published in the
Netherlands as Blijf de aarde trouw: Pleidooi voor een nietzscheaanse terrasofie, 2016 Henk Manschot
and Uitgeverij Vantilt, Nijmegen." | Includes bibliographical references and index.
Identifiers: LCCN 2020029972 | ISBN 9781350134393 (hb) | ISBN 9781350189423 (paperback) |
ISBN 9781350134409 (epdf) | ISBN 9781350134416 (ebook)
Subjects: LCSH: Nietzsche, Friedrich Wilhelm, 1844–1900. | Environmental sciences–Philosophy.
Classification: LCC GE40.M349513 2020 | DDC 193–dc23
LC record available at https://lccn.loc.gov/2020029972

ISBN: HB: 978-1-3501-3439-3
PB: 978-1-3501-8942-3
ePDF: 978-1-3501-3440-9
eBook: 978-1-3501-3441-6

Typeset by Deanta Global Publishing Services, Chennai, India

To find out more about our authors and books visit
www.bloomsbury.com and sign up for our newsletters.

This publication has been made possible with financial support
from the Dutch Foundation for Literature.

N ederlands
l etterenfonds
dutch foundation
for literature

Contents

Preface	vii
Acknowledgements	xv
Translator's note	xix
Abbreviations	xx

Part I Nietzsche's experimental lifestyle — 1

1. Nietzsche, animals and nature: A personal and philosophical voyage of discovery — 3
   - Nietzsche and the animals — 5
   - Nietzsche by the sea — 9
   - Nietzsche in the mountains — 15
2. Interacting with the natural world around us: On personal lifestyle and local culture — 21
   - Eating, fasting and diet — 24
   - Gardens and gardening — 27
   - Climate and cosmos — 34
3. *Thus Spoke Zarathustra*: A tragedy in four acts — 39
   - Act One: On the road with Zarathustra — 44
   - Act Two: The game of power — 47
   - Act Three: On dreams, time and eternity — 52
   - Act Four: The steps to the overman — 56
   - Recap: The great transformation — 61

Intermezzo: Gary Shapiro, *Nietzsche's Earth: Great Events, Great Politics* (2016) — 63
   - Three reflections — 63

Part II 'Terrasophy': Guidelines for a philosophy of the future — 71

4. Nietzsche and 'I': A cure for the lone individual — 79
   - Perception and expression — 81

	Experimental life	84
	Transforming ourselves	91
	A genealogical analysis of our own past	96
	A yes-saying morality	99
	Scenarios for the future	101
5	The place on earth: A revaluation of the local perspective	104
	Local culture: An ethnological perspective	107
	An ethnological view of European history	111
	Drawing upon the perspective of indigenous cultures	113
	Local ecological awareness in emerging economies	114
	The issue of urbanization	116
	The West and the local	117
	Two questions for further research	119
	Becoming indigenous: A philosophical reflection	122
6	Remain faithful to the earth: Towards an ecological cosmology	124
	A changed perspective on the earth	124
	The Anthropocene: The context of our time	126
	Nietzsche's earth	130
	Gaia	131
	Gaia and Nietzsche's earth: Three differences	132
	The overman	134
	The Anthropos in the Anthropocene	136
	The overman and Anthropos: An important difference	137
	On the way to 'the great health'	138
Afterword		140
	How to continue in Nietzsche's footsteps	140
	How to look at the future	140
	Should we make of the earth a city or a garden?	142
	A new sense of place: Locality, proximity and plurality	143
	The way to a radical humanism?	145
Notes		147
Bibliography		161
Index		172

Preface

'We are the first generation to feel the effect of climate change and the last generation who can do something about it.' That tweet from President Obama on 10 October 2014 was retweeted thousands of times within hours and provoked a torrent of responses worldwide. The facts on which he based his diagnosis are well known. Every day we are faced with a huge influx of information about climate change as the result of human activity, telling of drought, deforestation, falling fish stocks and rising sea levels. With exhausting regularity, scientists bring us new insights and predictions. We are all too well aware of the possibility that generations alive today might be the last. The future suddenly has a face, and a threatening one. But the power of Obama's statement lay in the fact that along with an alarming message he suggested the possibility of a solution – we must act now, without delay – thereby signalling hope and readiness. Nothing can stop us if we decide to do something. The responsibility lies with us. Obama's call for action has since been endorsed by many as defining the most important task of our times.

Succinct statements are powerful instruments of consciousness raising, but they inevitably ignore the complexity of the current crisis, and therefore prompt scepticism and irritation as well as approval and enthusiasm. We are only now beginning to acknowledge the toll this takes on us as individuals. French philosopher Bruno Latour describes our response as a feeling of disconnectedness. Reports about the climate crisis are so all-encompassing and dramatic that it is barely possible for us to comprehend them and take them in, he writes. We are unable to carry on with our ordinary lives as if nothing is happening, yet at the same time incapable of assimilating the new facts and adjusting our norms and values accordingly. Our disconnectedness results in uncertainty, confusion, anxiety and unease. A sense of urgency is beginning to take hold of people, a realization that something fundamental needs to change, but it is not immediately clear what that change should be – which only compounds the confusion.

Phenomena of this kind have been described by scientists as 'contrast experiences', which affect people so deeply that they cannot go on living in the ways to which they are accustomed. In psychology the term refers to events that profoundly alter a person's life, such as divorce, the diagnosis of a terminal illness or the death of a child. People come up against the boundaries of life as they knew it. They experience their own vulnerability and are overwhelmed by unfamiliar emotions and by issues that cannot be integrated into their lives.

Contrast experiences can sometimes be caused by social events that leave deep marks and disrupt daily life, and in this broader sense, the climate crisis fits the definition rather well.[1] I find the term illuminating because it relates to large-scale events happening in the outside world that impinge upon our lives.

A second characteristic of contrast experiences is that they raise questions touching upon our experience of meaning and purpose. Fear and uncertainty are combined with an unfamiliar sense of being vulnerable and dependent. No matter how serious the ecological crisis, it becomes real only when I allow it to affect me and set me thinking about how I live, what I do, what I could do differently and whether any small changes I make are significant in the larger process of life on earth. Questions may also arise about our experience of nature, about the superior and ignorant stance modern humans have adopted towards the natural world, about how we treat animals, about being mortal, or about what would be lost should our species die out. These questions refuse to take a firm shape or to fit into neat patterns, but they all connect the urgent need to alter something in our personal ways of living with issues concerning the meaning of life. Chinese philosopher Confucius is supposed to have said that each of us has two lives; the second begins when we realize we have only one. This insight is now acquiring a broader significance. Contrast experiences create space to seek new ideas and new values appropriate to the situation in which we find ourselves, and they connect our personal questions with the need for cultural change, for a 'revaluation of all values', as Nietzsche called it. The ecological crisis forces us to rethink the lives we have led up to now. In this book I focus on the nineteenth-century philosopher Friedrich Nietzsche, who found himself confronted with just such questions. We might regard him as having gone ahead of us in addressing them.

Of all the philosophers I know, Nietzsche is the most sensitive to buried tensions within a culture. He detected conflicts no one had ever addressed before, spotting hidden shifts and repressed feelings. In the late nineteenth century, he felt intuitively that in modern times something fundamental was amiss with the relationship between humans and the earth. 'The earth has a skin; and this skin has diseases. One of these diseases for example is called "human being"' (Z, II, 'On Great Events') his alter ego Zarathustra says. He followed up his diagnosis with an appeal: 'Remain faithful to the earth' (Z, I, 'On the Bestowing of Virtue', §3'). I discovered only gradually what Nietzsche was referring to with his diagnosis and his injunction. Together they turn out to be the briefest possible summary of the philosophical project to which Nietzsche devoted his life.

He gave himself a triple task. First, he was determined to adopt a lifestyle that would bring him into closer contact with nature. Secondly, he wanted to develop a new philosophy in which the earth would be central, rather than humankind. Thirdly, he intended to express his growing criticism of modern culture, which had made human beings the basis for all attribution of value and extended that principle into all domains of life. Each of these three points requires elucidation, which I will give shortly, but together they make clear why, after seeking a philosopher who could help me to gain a bit of distance from the daily upheaval of the ecological crisis, I found myself turning to Nietzsche.

The era in which Nietzsche lived, the second half of the nineteenth century, was an unsettled time in Europe. Everything was in ferment, and political and military conflicts raged, including the Franco–Prussian War. The Industrial Revolution had set workers and the bourgeoisie against one another, each side supported by ideologies of its own, and tensions had arisen between the new sciences, with their rationalization of culture and religions, and the faith in god that continued to shape daily life. Art lost its moorings, painting and sculpture in particular. There was little evidence in those days that ecological problems would arise as a result of modernization – even for Nietzsche that must initially have been no more than a vague suspicion – but a radical change in lifestyle sparked new ideas. In about 1880 Nietzsche left the university in Basel where he had worked as a professor for ten years, opting instead for a life that would bring him into daily contact with nature. At the same time he started to put into practice a new way of philosophizing. Abandoning the detached thinking of academic philosophers, he defended the view that any philosophy worthy of the name must be tested out in the philosopher's own life. In his view philosophers prove the correctness of their ideas by living according to their own dictates, an art not taught in the universities. The relationship between his life and his thought became an important aspect of Nietzsche's new approach.

Eager to pay due attention to this dimension in writing about Nietzsche's philosophy, I tried to get as close as possible to his new way of life. All the biographies agree that between 1879 and 1889, the period in which he wrote his most important books, Nietzsche spent the summer months in the Swiss Alps around Sils Maria and took long walks, which lasted at least six hours a day. In the winter, when it was too cold for him in the Alps, he stayed on the Mediterranean coast and there adhered to a similar daily rhythm. The impact of his intensive outdoor life on his thinking has not been properly addressed by his biographers. I wanted to find out more about his daily walks, about how the climbs and descents, high and low altitudes, cold and warmth set him thinking about what happiness is, not as an abstract notion but as an experience. For the writing of this book I therefore travelled in Nietzsche's footsteps. I withdrew to the Alps for ten weeks and afterwards spent several weeks on the Mediterranean coast. I walked a lot, with his writings in my pocket, experiencing for myself what it does to you, to your thinking and to your life, when the horizon is formed by the mountains or the great expanse of the sea rather than the neighbours across the street, when the pace of life is that of steady walking, of sunrise and sunset, of the tides, rather than the speed of the car or the internet. I wanted to feel how nature became the companion Nietzsche needed if he was to move forwards in his thinking or allow an idea to rise slowly within him that he did not yet dare admit in its full extent, such as the belief that human beings are the living dead – modern human beings, that is. It takes courage to think new thoughts, he writes repeatedly, adding that it also means being alone a great deal.

It puzzled me that Nietzsche tended to express his ideas and thoughts best when he thought in visual rather than verbal language. How did he avail himself of the linguistic register of nature, of landscapes, of different climates and above all of animals? Of all modern philosophers, it is Nietzsche who mentions the largest number of animals in his writing, no fewer than 120 different species, and they are astonishingly diverse. The time Nietzsche spent outdoors and his perpetual search for vivid language were indispensable in his quest for a new philosophy of the earth. Animals were vital sources of nourishment for his philosophical thinking, which in turn inspired changes to his lifestyle. Not only did he walk a great deal, he experimented too, especially with his diet, contemplating the ways people feed themselves, and he sought out the places on earth that enabled him to think and to live. His lifestyle and his style of writing and thinking continually interpenetrate.

Having taken his personal experiences as his starting point, Nietzsche then set these against his experience of modern culture. He became intensely frustrated by the contemporary way of living, and the morality and religion that went with it. He felt that modern culture 'degenerated' people by separating them from the earth, and 'spiritualized' them in the worst sense of the word. These were two sides of the same coin: the more modern a life, the further it was from nature. His book *Thus Spoke Zarathustra*, which contains Nietzsche's plea for a lifestyle that he calls not ecological but 'faithful to the earth', was written between 1882 and 1884. He takes the reader on a voyage of discovery. It becomes clear as we read that the elevation of 'man' to the measure of all things alienates us from our bond with the earth and cuts us off from the natural processes of life. The other books Nietzsche wrote in the early 1880s include the collections of aphorisms *Human, All Too Human*, *Daybreak* and *The Gay Science*. He also wrote many notes, which have come down to us. But he left no room for doubt that *Thus Spoke Zarathustra* was the book in which he had truly recorded his vision of the earth. For that reason it is the book I have taken as a guide, although the same subject features in many of his other writings. In retrospect we can see that the beginnings of what we would now refer to as his ecological consciousness were present earlier, in what he calls the books of his wandering years, in unexpected intuitions that punctuate his work, in observations that surprise even him, in meditative reflections, and in his interest in evolution and biology.

All this became clear to me only gradually. Nietzsche's criticism of our modern culture and lifestyle increasingly takes the relationship between humans and the earth as its main focus. Naturally it was hard for him to free himself from the patterns of thinking within which he had grown up. 'We are entwined in an austere shirt of duty,' he writes (BGE §226)[2]. But gradually, through a process of trial and error, he found the words by means of which he could extricate himself from that tight garment. Nietzsche is known above all as a philosopher who saw life as driven by the 'will to power'. We also know him as the philosopher who concluded that 'God is dead' and helped to fuel the nihilism that resulted. But ever since I discovered Nietzsche's ecological concerns, it has been my aim to explore his attempts to get closer to nature, to investigate what he meant by taking as his guiding principle 'remain faithful to the earth', and in doing so to explore how he might inspire us in our own time.

Gary Shapiro, in his book *Nietzsche's Earth: Great Events, Great Politics*, has recently reconstructed the political purport of Nietzsche's philosophy of the earth.[3] Nietzsche's scepticism regarding the dominant culture was accompanied by an intense desire to look at the earth in a completely different way from that presented to him by modern culture, and to create openings for a life that in his view was worthy of the name.[4] Zarathustra addresses all this, raising fundamental questions about morality, religion and art. Nietzsche tells us that if we as modern humans want to get back into contact with the earth, we must dare to make the meaning and significance of a 'grounded' human life our central concern. The future of our species depends on it. Nietzsche therefore examines modernity's hostility to nature and the core values upon which it is based. His aim is to produce a philosophy that will once again make the universe, and especially the earth, a guide to humanity, a new philosophy of the earth that goes beyond both classical and modern cosmology.

In my contemplation of the relationship between humankind and the earth in the present day, I have made Nietzsche my companion, addressing to him questions of my own. Of course other philosophers will join the conversation. In examining his work and his lifestyle I focus on three themes that I found surprising and new. In the first part of the book I follow Nietzsche closely and highlight some of the things he has taught me. This produces a framework for my own reflections, which I address in Part II. Taken together, these three themes can make a significant contribution to a contemporary philosophy of ecology.

The first theme is an obvious one: a new philosophy of the earth demands of modern humans that they reconnect with the natural world around them in their daily lives. We soon realize, however, how difficult this can be. In place of philosophy as an academic study, Nietzsche advocates a personal engagement in which living and thinking are continually connected. What does restoring contact with nature mean for our lives and what new kinds of thought and reflection emerge from it? In an attempt to answer this question I follow Nietzsche and show how, step by step, his own way of thinking changed. Instead of the academic philosopher he once was, he became a walker, writing aphorisms about what inspired him and defining life as one great experiment.

The second theme concerns our relationship with the place where we live. In his ecological philosophy, Nietzsche attributes a vital significance to this relationship. He even makes it central to his concept of culture as mediating between people and their own patch of earth. Among the questions he asks are: How does this cultural interaction work? How does the land occupied by a tribe, a village, a city or a people speak through its customs and traditions? What rites and festivities, commandments and prohibitions, stories, costumes and masks, fashions and make-up, eating habits or painting and sculpture does a community create about its place on earth and how? In what ways do people care for and cultivate their environment, so that it acquires greater significance for them? Finally, why did modernity break with the pluriformity of these traditions? Nietzsche believed that for centuries earthly life featured a varied, multicoloured and always surprising patchwork of social creativity. Culture is fundamentally pluralist in nature, and in Nietzsche's vision it must become so again in the future. He was deeply troubled by the fact that modernization, with its universal moral rights and duties, encourages uniformity. In an era of globalization his ideas have gained new relevance, as the pressure exerted by the neoliberal global economy on local living conditions proves increasingly incompatible with sustainable lifestyles.

My third theme is derived from Nietzsche's reaction to the modern concept of the earth as a lifeless heavenly body. By freeing the earth from the image we have had of it since Galileo and Newton, he attempts to look at life on earth with fresh eyes, seeing our planet as a living entity, as awe-inspiring, as the other, the larger – as a reality that cannot be appropriated or possessed. In his early notebooks he explores all the many ramifications of this fresh insight. By freeing the earth from its modern depictions, Nietzsche enters into a new, open connection with it, from which emerges a figure of great vitality that Nietzsche calls the 'free spirit'.

In our own day we realize how complex and intense is the connection between the various ecosystems on planet earth. They keep each other in balance as part of one immense system, but that balance is now under threat. Some have concluded that many generations to come will confront the resulting imbalance, a diagnosis that leads European philosophers including Bruno Latour, Isabelle Stengers and Peter Sloterdijk, from whom we shall hear more in the final chapter of this book, to the conclusion that it is high time for us to wake from the modernist dream and face the seriousness and urgency of the situation we are in. Sloterdijk agrees with philosopher Hans Jonas that the only reality that might yet unite all people and impose a categorical imperative is the earth. Jonas is remembered for his ecological imperative: 'Act in such a way that the consequences of your actions are compatible with the permanence of true human life on earth.'[5] The time is ripe for an ethics of ecological solidarity and a world citizenship that gives expression to it. We are coming to realize how strongly bound together we are from an ecological point of view, but our moral development is lagging behind. A comprehensive cultural shift is needed, placing life on earth as a whole, rather than merely the interests of our own species, at the centre of our concerns.

Acknowledgements

During my first sabbatical, in the early 1980s, I roamed the mountains of Nepal for six weeks. I had two books in my rucksack, *Siddhartha* by Herman Hesse and *Thus Spoke Zarathustra* by Friedrich Nietzsche. It was a real sabbatical, far away from the goings on of the university world that could be fairly hectic even in those years. At the end of the trip one thing was clear to me: I wanted to do something like it again one day. But I had to wait until after retirement to be able to follow in Nietzsche's footsteps, in the Alps and along the Mediterranean coast. Neither of those walks would have been possible without the hospitality of many friends. My French in-laws' apartment in the Swiss Alps became a precious place, where I could combine walking and writing. In 2013 I spent several weeks hiking in the Alps on the Italian side, and there I had a beautiful Walser 'hut' to stay in, built according to old German methods. A few of these buildings can still be found in the region around the Monte Rosa. The spring of 2013 turned out to be colder than expected, and on those frosty days Raymond and Blanca, and their extraordinary children Beaudelot, Quinteyn and Madalie, were a heart-warming source of pleasure, inspiration and friendship. Their dedication to the 'vita pura' – their term for a style of life close to nature that they are developing there, along with the unforgettable Mr Luca, his vegetable plot and his goats – gave me the feeling that my intuitions had put me on the right track. Because I prefer warmth to cold and because Nietzsche chose to live on the mild coasts of the Mediterranean in winter, I wanted to make that part of his experience my own too. An offer from another branch of my wife's family enabled me to do so. I spent a solitary and peaceful March–April season on the rugged coast of Spain. The light and warmth of the place and the endless view out over the sea will always be connected in my mind with the sunniest passages of the book. My heartfelt thanks go to the Vincenots, Landgraafs and Jannots.

Writing this book involved alternating between periods of being alone, of concentration and seclusion, and periods of conversation and discussion with friends, colleagues and students. Their interest, suggestions and comments helped me on my way again and again. Joep Dohmen gave me the idea of starting this project. He recommended books about Nietzsche and followed developments from start to finish with his sometimes critical, sometimes surprised but always encouraging view. Only very gradually did I gain a grip on the material. The ability to test the progress of my ideas annually at the International Summer School organized by the Kosmopolis Institute of the University of Humanistic Studies helped immensely. I would like to thank the students for their enthusiasm and the staff for their sincere commitment. My dear friend and colleague Caroline Suransky, coordinator and source of inspiration for the Summer School, experienced the entire process from close proximity. We had many conversations on the subject of the book. She annotated several versions of the text and at times of doubt she always encouraged me not to give up. Another exceptional colleague at the Summer School, Sitharamam Kakarala from India, became an enthusiastic sparring partner. Ram is an unerring expert on computers, a fervent practitioner of Ayurvedic medicine and a great lover of Indian music, which he shared with us. But it was above all his sharp observations and suggestions that left their mark at several places in the book. With Zainal Bagir from Indonesia I share an interest in developments in the field of religion and ecology, especially within Islam. J.C. van de Merwe, director of the Institute for Reconciliation and Social Justice at the University of Bloemfontein in South Africa, offered to develop the themes of the book further at his institute. I also received a great deal of support from friends and colleagues at the University of Humanistic Studies. I am grateful to Hans Alma, chair of the research group Globalisation and Dialogue Studies, for allowing me to remain attached to the university via her chair. I had stimulating conversations about the importance of a terrasophical perspective for the study of humanism with my old student, faithful friend and colleague Laurens ten Kate, with Fernando Suarez Muller and with Renske van Lierop. Koo van der Wal, Martijn Rozing and Eric van der Vet supported me with their advice, while Patrick Vlug carefully worked through the final version of the text and corrected it meticulously. Then there were the people I have worked with for years. Ilse Bulhof, who has written about Nietzsche herself, pointed out to me the link with Buddhist ideas, and my old friend Connie Lips followed my progress closely and made the subject of how to put philosophy into practice a recurring theme in our conversations.

Acknowledgements

A special word of thanks goes to Harry Kunneman. Walking along the River Lek near Culemborg and during our delightful meals, which he often cooked himself, we went through the main topics of the book on several occasions. Harry managed time and again to make me look at my ideas in a new way. He was working on a book on a related theme at the same time, so we were able to discuss many issues. In the final phase he and Caroline were crucial. Without their responses this would have been a different book. Paul van Tongeren, who offered to read the final manuscript and responded warmly and positively, opened the door to publishing house Vantilt. I am grateful to Henk Hoeks and Mayke van Dieten for carefully editing the text and for the meticulous attention my footnotes required; it was a thoroughly pleasant collaboration. My thanks to Marc Beerens for choosing to publish the book. Along with Caroline Suransky I went in search of an English-language publisher. To my great pleasure, Bloomsbury proved willing to publish a revised version of the book. The collaboration with Liza Thomson and Lisa Goodrum, which introduced me to the many rules of the publication game, was cordial. Fortunately Liz Waters was prepared to translate the manuscript. Her reputation as a translator is well known. It was an immense pleasure to discover how much care and attention she had paid to the text, with great sensitivity to the philosophical and literary subtleties of Nietzsche's thinking and his significance in our own day. It was a unique experience to work with her on the English edition.

The people with whom I share my life have been involved in this project in a way all their own. That we have lived for so long on the Frans Halsstraat in Utrecht together with other people and an exceptional cat not only made life agreeable, it ensured that I could easily spend long periods away. For that I thank Dineke Admiraal, my sister Maria and my wife Agnes Vincenot. I admire the way in which our sons Menno and Tim, along with Emma and Hubertine and their children Olli, Yannic, Marilou, Nora and Vesper have devoted themselves to helping to create a more sustainable and humane world, and how they give shape to these values in their work and in their lives. You were often in my thoughts as I wrote this book and I hope that you – and even more so the little ones – will find inspiration in it for the future. I was able to talk over the subject of the book with Menno and Tim, and they relieved me of the organizational rigmarole involved in its launch. I sincerely hope that we can continue our collaboration for years to come. Agnes was the most closely involved with the whole project. It was she who took me into the mountains for the first time, almost fifty years ago, and transformed me from a son of the lowlands into a mountain person. She ensured that every year, in practice, our daily life became a little more 'green'. She imparted to me her love of gardening, for example, and still dreams that one day we will take a long-distance walk with a donkey. She gave me the freedom to carry out my project however I wished. She was interested in my writing and understands the art of weighing every word. She likes to place a question mark beside every concept, and although I did not by any means adopt all her suggestions, she continued to surround me with her love, tenderness and loyalty. It is a great joy to see how the past few years have brought us closer together.

Extension of the copyright page

Excerpts from *The Gay Science* by Friedrich Nietzsche, translated by Walter Kaufman, copyright ©1974 by Penguin Random House LLC. Used by permission of Random House, an imprint and division of Penguin Random House LLC. All rights reserved.

Excerpts from *Thus Spoke Zarathustra* by Friedrich Nietzsche, translated by Adrian Del Caro, copyright © 2006 by Cambridge University Press. Used by permission of Cambridge University Press.

Permission has been granted by UNESCO to use excerpts from the Universal Declaration on Cultural Diversity (2001)©UNESCO http://portal.unesco.org/en/ev.php   URL_ID=13179&URL_DO=DO_TOPIC&URL_SECTION=201.html

Translator's note

Since no comprehensive English translation exists of the many notes and letters left by Nietzsche upon his death (referred to in German as *Nachgelassene Fragmente*), his unpublished work has been rendered into English by the translator, unless otherwise stated. References to this work (beginning NF, or in the case of the letters BVN) relate to the eKGWB, the *Digitale Kritische Gesamtausgabe Werke und Briefe* or 'digital critical edition of the complete works and letters', based on the critical text by G. Colli and M. Montinari (see KSA, KSB, KGW and BVN). This digital edition, published by Nietzsche Source, is freely available via http://doc.nietzschesource.org/en/ekgwb.

Abbreviations

A list of works by Friedrich Nietzsche, with standard abbreviations. Publication details are given for the translated editions cited in the text.

A   *The Antichrist / Der Antichrist* (1895)

ADHL   *On the Advantage and Disadvantage of History for Life*, translated by Peter Preuss, Cambridge/Indianapolis: Hackett, 1980. Translation of *Unzeitgemässe Betrachtungen. Zweites Stück: Vom Nutzen und Nachtheil der Historie für das Leben* (1874)

BGE   *Beyond Good and Evil: Prelude to a Philosophy of the Future*, translated by R.J. Hollingdale, London: Penguin Books, 1973, rev. 1990. Translation based on both the first edition and the standard edition of *Jenseits von Gut und Böse: Vorspiel einer Philosophie der Zukunft* (1886/1894)

BT   *The Birth of Tragedy from the Spirit of Music / Die Geburt der Tragödie aus dem Geiste der Musik* (1872)

BVN   *Nietzsche Briefwechsel. Kritische Gesamtausgabe*, edited by Giorgio Colli and Mazzino Montinari, Berlin/New York: de Gruyter (1975) (see Translator's note above)

D   *Daybreak: Thoughts on the Prejudices of Morality*, translated by R.J. Hollingdale, edited by Maudemarie Clark and Brian Leiter, Cambridge: Cambridge UP, 1997. Translation of *Morgenröte. Gedanken über die moralischen Vorurteile* (1881)

EH   *Ecce Homo: How One Becomes What One Is*, translated by R.J. Hollingdale, London: Penguin Books, 1979/1992. Translation of *Ecce Homo. Wie man wird, was man ist* (written 1888, first published 1908)

GM	*On the Genealogy of Morality: A Polemic*, translated by Carol Diethe, edited by Keith Ansell-Pearson, Cambridge: Cambridge UP, 1994/2007. Translation of *Zur Genealogie der Moral: Eine Streitschrift* (1887)
GS	*The Gay Science: With a Prelude in Rhymes and an Appendix of Songs*, translated by Walter Kaufmann, New York: Random House, 1974. Translation based on the second edition of *Die fröhliche Wissenschaft* (1887)
HA	*Human, All Too Human: A Book for Free Spirits*, translated by R.J. Hollingdale, Cambridge: Cambridge UP, 1986. Translation of *Menschliches, Allzumenschliches. Ein Buch für freie Geister* (1878)
KSA	*Sämtliche Werke. Kritische Studienausgabe*, 2nd edition, edited by Giorgio Colli and Mazzino Montinari, 15 vols., Berlin and New York: de Gruyter; Munich: dtv (1988)
KSB	*Sämtliche Briefe. Kritische Studienausgabe*, edited by Giorgio Colli and Mazzino Montinari, 8 vols., Berlin and New York: de Gruyter; Munich: dtv (1986)
KGW	*Kritische Gesamtausgabe Werke*, edited by Giorgo Colli and Mazzino Montinari. Berlin/New York: de Gruyter (1975)
NF	*Nachgelassene Fragmente* (see Translator's note above)
TI	*Twilight of the Idols, or, How to Philosophize with a Hammer / Götzen-Dämmerung, oder, Wie man mit dem Hammer philosophiert* (1889)
TL	*On Truth and Lies in a Nonmoral Sense / Über Wahrheit und Lüge im aussermoralischen Sinne* (1873)
UM	*Untimely Meditations / Unzeitgemässe Betrachtungen* (1873–6)
WS	*The Wanderer and His Shadow*, translated by R.J. Hollingdale, Cambridge: Cambridge UP, 1986, published as part of *Human, All Too Human* (see above). Translation of *Der Wanderer und sein Schatten* (1880)

Z   *Thus Spoke Zarathustra: A Book for All and None*, translated by Adrian Del Caro, edited by Adrian Del Caro and Robert Pippin, Cambridge: Cambridge UP, 2006. Translation of *Also Sprach Zarathustra. Ein Buch für Alle und Keinen* (1883–5)

Part I

Nietzsche's experimental lifestyle

1

Nietzsche, animals and nature
A personal and philosophical voyage of discovery

Around 1880 Nietzsche set off in an entirely new direction in life. He was thirty-six years old and had resigned from his position as a professor of classical philology in Basel to seek peace and quiet. He went looking for a place where he could feel healthy and be able to think and write. For the summer months he found it in the Swiss Alps near Sils Maria, 'at an altitude of six thousand feet above sea level'.[1] Almost every summer after that he spent several months in a small boarding house built up against a rock face. He needed only to step out of the door and he was in the mountains. In winter he preferred the mild climate of southern Europe, although he did not find there one specific place in which to settle. He stayed mainly in or near Genoa, Nice or Venice. One thing all three places have in common is their coastal location. With a view of the sea and the mountains, this was the landscape in which Nietzsche took his daily walks, lasting up to seven hours a day. Along the way he made brief notes about ideas on all kinds of subjects, which he worked up into short passages of text in the evenings or in the days that followed.

 He chose to walk alone. His repeated forays into the mountains or along the rocky Mediterranean coast put their stamp on his life and determined the rhythm of his days. Nature was his faithful companion and silent witness to his moods and ruminations, and in it he encountered an endless stream of impressions, associations and fantasies. He experienced his new life of peace and fresh air as a profound liberation. He felt his whole body come alive. Powerful emotions took hold of him, primeval passions that existed both in him and in the natural world around him. A desire to come into contact more and more deeply with 'le sens immédiat de la vie', with life as an immediate and unmediated experience, is how Italian philosopher Giorgio Colli, editor of the first scholarly edition of Nietzsche's work, described the most basic motivation that inspired Nietzsche from this time onwards. His philosophizing changed radically as a result, in both form and content. Not just thinking and speaking but singing and dancing and the poems of 'Prince Vogelfrei' all became part of it, a whirlwind of associations and reflections that seem born out of new experiences.

The first fruits of this walking philosophy were collected in books including *Human All Too Human*, *The Dawn*, *The Wanderer and His Shadow* and *The Gay Science*. The title of the latter refers to the knowledge of the troubadours. In a revised preface, added after a few years to a second edition of *The Gay Science*, he characterized the experiences of those early years of wandering as a process of healing for body and mind. Out of the sick and degenerate situation in which he had found himself, 'one returns newborn, having shed one's skin, more ticklish and malicious, with a more delicate taste for joy, with a tenderer tongue for all good things, with merrier senses, with a second dangerous innocence in joy, more childlike and yet a hundred times subtler than one has ever been before' (GS, Preface, §4). It was clearly a truly exceptional experience of healing of both mind and body, and just how powerful it must have been becomes even clearer when we realize that Nietzsche had been feeling sick, in every sense, of the culture in which he lived. This expressed itself in feelings of contempt and an almost physical aversion to his own era, as well as rage and powerlessness, sarcasm, a bodily sense of not being at home anywhere, migraine, eating disorders and difficulty sleeping. His sufferings reverberate throughout the text, which continues,

> How repulsive pleasure is now, that crude, musty, brown pleasure as it is understood by those who like pleasure, our uneducated people, our rich people, and our rulers! How maliciously we listen now to the big county-fair boom-boom with which the 'educated' person and city dweller today permits art, books, and music to rape him and provide 'spiritual pleasures' – with the aid of spirituous liquors! How the theatrical scream of passion now hurts our ears, how strange to our taste the whole romantic uproar and tumult of the senses have become, which the educated mob loves, and all its aspirations after the elevated, inflated, and exaggerated! (GS, Preface, §4)

Nietzsche felt forced to give up the life into which he had settled, to terminate the professional career that had made him a respected figure in the cultural world of his day. He was determined to live differently, closer to nature. But it is typical of Nietzsche that his personal process of physical and mental healing was from the start tightly interwoven with questions about culture. He applied the labels 'sick' and 'healthy' to cultures as well as to individuals, and he believed cultures could make people ill or extinguish their joy in living. Might they also help to promote high spirits? If so, how? A new skin, a softer tongue, a more delicate taste – primary physical experiences and sensations were his first guides on the way to the 'great health' he was seeking. Animals had a vital part to play in that search.

Nietzsche and the animals

During his walks, Nietzsche must have developed a special bond with animals and indulged the emotions and fantasies that animals foster in human beings. Yet it was not immediately clear how they were to figure in his life. Although I have been reading Nietzsche's work for many years, it came as a complete surprise to discover that of all modern philosophers it is he who discusses animals most often and in the greatest variety. It came home to me only when I discovered the book *Bestiaire de Friedrich Nietzsche* (2011), in which François Brémondy lists all the living creatures mentioned in Nietzsche's books in alphabetical order, complete with quotations from the passages in which they feature. There are well over a hundred of them.[2]

They fall into two categories. First, there are those Nietzsche must have come upon on his walks and that caught his eye, including a lot of birds, naturally: swallows, pigeons, seagulls, sparrows, probably the eagle, and in his imagination the albatross, which along with the eagle made the greatest impression on him of all. Then there were the large herbivores and herd animals, the cows and sheep, the horses and donkeys, as well as the dog, cat, goose, peacock, snake, frog, and a menagerie of butterflies, flies and hornets, worms and ants. The other category is made up of animals that he cannot have seen himself but that very much spoke to his imagination and his state of mind: the camel and the lion, the tiger and the bear, the wolf and the llama, the rhinoceros and the hyena, the crocodile and the rattlesnake, not forgetting the ape. Many of these were to have a considerable impact on his ideas about inversion and transformation.

What functions do animals go on to fulfil in Nietzsche's philosophy? The following passage contains an early hint.

> Moreover my eyesight is poor and my imagination (whether dreaming or awake) is accustomed to a great deal and regards a great deal as possible that others would not always accept as such. – I fly in dreams, I know it is my privilege, I do not recall a single situation in dreams when I was unable to fly. To execute every sort of curve and angle with a light impulse, a flying mathematics – that is so distinct a happiness that it has permanently suffused my basic sense of happiness. When I'm in a very good mood, I glide about that way continually, high and low as I see fit, the one without effort, the other without haughtiness and abasement. 'Upswing', as many describe this, is for me too muscular and violent. (NF-1881, 15 [60])

The bird is a major figure in Nietzsche's writings. Zarathustra, Nietzsche's alter ego, says, 'If ever I spread silent skies above me and flew into my own sky with my own wings . . .,' he also talks of 'my freedom's bird-wisdom' (Z, 'The Seven Seals, §7'). Nietzsche feels akin above all to the big birds that hover, including the albatross, king among birds, and the condor of the Andes (NF-1883, 8[2]), able at the greatest of heights to hang almost motionless in the air for more than an hour, riding the air currents. On the wings of these birds Nietzsche explores the landscape and his feelings of appreciation and disdain, experiencing bodily the energy that is released when a person can be free as a bird, nourishing a desire for all things light, graceful and high. Birds teach Zarathustra to sing and urge him to stop talking for once. Of all the animals it is the birds that inspire in him the most beautiful words with which to capture intense moments of happiness and deep desires.

Animals accompany Nietzsche in his search, understand his fears and joys, bring him 'food' in times of need and sing along with him in moments of joy. We find them doing all these things in his book *Thus Spoke Zarathustra*, a dramatized frame story about Zarathustra who withdraws to the mountains where, as a hermit, he experiences the great transformation that occupies Nietzsche's thoughts. The animals 'did not leave his side day and night, unless the eagle flew out to fetch food' (Z, III, 'The Convalescent, §2'). Here is Nietzsche the poet, the prophet–poet as Zarathustra calls himself. Here animals can do whatever poets and storytellers make them do. The lion can laugh and mice dance on the table. Nietzsche pulls out all the stops and takes inspiration in doing so from literature and from animal philosophy. The eagle that 'flew out' brings so much back that

> Eventually Zarathustra lay among yellow and red berries, grapes, red apples, aromatic herbs and pine cones. At his feet, however, two lambs were spread out, which the eagle with difficulty had taken as prey from their shepherds. Finally, after seven days, Zarathustra sat up on his bed, picked up one of the red apples, smelled it, and found its aroma lovely. Then his animals believed the time had come to speak with him. (Z, III, 'The Convalescent, §2')

Animals keep humans company in many stories. That in itself is nothing new. What strikes me here is the richness of the colours and smells of nature, and the warmth with which Nietzsche writes about Zarathustra's animals. There is a sharp contrast between Zarathustra's contact with animals and the way he interacts with people – as if Zarathustra belongs to their world; as if he were one of them.

Animals help Nietzsche to detect intense urges that we do not easily accept as part of ourselves, feelings Nietzsche believed were appropriate to hunter and prey. From the very start he is fascinated by the phenomenon of violence in nature and wrestles with its significance. How do life and violence go together? What impulses show themselves in violent and cruel behaviour? What do we feel if we empathize with such behaviour, indulge it, allow ourselves to be swept along by it? Nietzsche is fascinated by the superior beauty of the predator, by its great power, and he comes to see that compassion, which lies at the core of morality, holds no value for it. He repeatedly states that morality has no part to play in the interaction between predator and prey, indeed must play no part in it. After one of his walks he noted, 'In nature there is no partiality for the living or against death. If something living does not survive, nothing fails in its purpose' (NF-1881, 12[111]). The place of violence and aggression in life, including human life, became one of Nietzsche's central themes. In his subject matter the centre of gravity eventually shifts to the moral perspective. What kind of violence contributes to life? With such questions in mind he consulted the latest scholarship, becoming involved in debates about evolution, investigating the new science of biology, in which organic life is explained as a great interplay of forces and counterforces. Later, in *On the Genealogy of Morals*, he diagnosed the coming into being of human morality as one long, progressive development of increasingly complex schooling, of attempts to discipline the primal force that the predator, which cannot be disciplined, inflicts in its natural beauty and splendour.

These are some of the ways in which animals appear in Nietzsche's writing. It would be possible to identify others, but I do not intend my account to be exhaustive, and anyhow there is something artificial about picking a text apart in this way. Nevertheless, it has helped me in my efforts to discover how important perception and empathy were for Nietzsche in his discovery of nature, in restoring his sense of a connection with life, of *le sens immédiat de la vie*. The overwhelming vitality of a life lived outdoors resonates with a rich variety of poetry and philosophy that came from inside him.

Almost all the literature about Nietzsche quotes a sentence in which he characterizes humanity as 'the animal whose nature has not yet been fixed' (BGE, III, §62), words usually explained as a theoretical assertion. Nietzsche is often assumed to have meant that all animals besides humans have fixed patterns of instinct. Or that the human being is not an animal among animals but differs from them categorically. I suspect that this particular pithy statement was preceded by a perception. Nietzsche was so good at imagining what it was like to be an animal that during one of his walks it must have occurred to him to ask: If I were to create myself, what would I want to be like? In what combination of strengths and skills would I wish to train myself?

Animals occupied a key role in Nietzsche's new life, but in his writings he developed his own particular way of associating with them. We need to become familiar with it, even to go along with it, if we are to understand him. His animals want to put the people who have shut themselves up in their human world back on track towards true happiness. On a long walk in the mountains we might suddenly come upon a small group of cows that stand together and give us a good long look. Such an experience was one Nietzsche would remember. It turns up in *Thus Spoke Zarathustra*.

> When Zarathustra had left the ugliest human being, he was freezing and he felt lonely; after all, so much that was cold and lonely went through his mind, to the point where even his limbs grew colder because of it. But as he climbed further and further, up, down, now past green meadows, but then also across wild stony deposits where previously an impatient brook might have laid itself to bed, then all at once his mood became warmer and more cordial. 'What happened to me?' he asked himself, 'something warm and lively refreshes me, something that must be close to me. . . .' But when he

peered about himself and searched for the comforters of his solitude, oddly enough, it was cows huddled together on a knoll; their nearness and smell had warmed his heart. Now these cows seemed engrossed in listening to someone speaking, and they paid no attention to the one who approached them. But when Zarathustra was quite near them he heard clearly how a human voice spoke from the midst of the cows; and evidently they had all turned their heads toward the speaker.... 'What are you seeking here?' cried Zarathustra, astonished. 'What am I seeking here?' he answered: 'The same thing you seek, you trouble maker! Namely happiness on earth. But for that I want to learn from these cows.' (Z, IV, 'The Voluntary Beggar')

Where did Nietzsche encounter his animals? Why was it that he found among them his most faithful companions, rather than among people? To understand this we need to go a few years back in time.

Nietzsche by the sea

> 'So is there in the whole world now a single person who, like me, sits beside the sea and . . .' (NF-1881, 12[113])

Nietzsche's decision to lead a different life did not come out of the blue. The years that preceded it were far from happy. In 1876, four years before the start of his period of wandering, when he was still a full-time professor, he took a trip to Italy. It was his first journey to southern Europe. Greatly troubled by migraine and other ailments, mentally exhausted and disappointed by Wagner's cultural projects, he accepted an invitation from a good friend to go and stay with her in Sorrento, near Naples, to recuperate. He told her he would bring two friends with him. According to Paolo D'Iorio, the stay in Sorrento brought about the first fundamental transformation in Nietzsche's life and thought. We can follow the events of those few months almost day by day, since in D'Iorio's 2012 book *Le voyage de Nietzsche à Sorrente* (published in English in 2016 as *Nietzsche's Journey to Sorrento*), the author reconstructs them by drawing upon Nietzsche's own notes, diaries and letters, and those of the other guests. He summarizes Nietzsche's stay, not without pathos, as follows:

> In Sorrento, in the large bedroom on the third floor of the Villa Rubinacci, which looks out on an orange grove and, farther off, by the sea, onto Mount

> Vesuvius and the islands of the Gulf of Naples; in the luminous autumn afternoons, silent and orange-scented, still pervaded by midday sun and sea salt; during the evenings of reading aloud with friends, or during the day-trips to Capri or to the carnival in Naples; on walks through the little villages that extend along one of the most beautiful gulfs in the world, on this earth where the ancients believed they heard sirens; during the mornings spent writing the first aphorisms of his life, of which the drafts still carry the name *Sorrentiner Papiere* today, Nietzsche decides to become a philosopher. (D'Iorio, 2016, p. 4)

All the ingredients that made this stay so special come together here. In the Villa Rubinacci, just outside the village, hostess Malwida von Meysenbug, along with her *dame de chambre* Trina, received Nietzsche and his friends Paul Ree and Albert Brenner on 27 October 1876. They all stayed there together for about six months and soon developed a regular daily routine and a distinctive ambiance. 'Our life in Sorrento organized itself very comfortably,' writes Malwida in one of her letters.

> In the morning we were never together; everyone attended to his own occupations in total freedom. The midday meal was the first to reunite us, and sometimes in the afternoon we would take a stroll together through the enchanting surroundings, among the gardens of orange and lemon trees as tall as our apple and pear trees and whose branches, covered in golden fruit, bent over the garden walls and cast their shadows along the path; or we would climb up gently sloping hills and pass by farms where lovely girls were dancing the tarantella. . . . Often, we would take longer excursions, riding on donkeys, which are reserved there for mountain paths, and our laughter and merriment on those occasions knew no bounds; the young Brenner especially, with his awkward, schoolboyish manner and long legs that nearly trotted alongside those of the donkey, was the target of many good-natured jokes. In the evening, we reconvened for dinner and then in the sitting room, for animated conversation and communal readings. (D'Iorio, 2016, p. 37)

She tells us that Nietzsche rode horses and sometimes went out walking on his own. The 'monastery of free spirits', Nietzsche called the villa, and he confided in Malwida that he had never felt so good before in his life. 'He is beginning to sense what health is,' Malwida wrote to a friend (D'Iorio, 2016, p. 38).

They read an astonishing amount together: Voltaire and Diderot at first, soon followed by Thucydides, Plato, Herodotus, the New Testament, Goethe, Mainlander, Spir, Burckhardt, Ranke, Michelet, Daudet, Calderón, Cervantes and Turgenev. Even this list is not exhaustive. Reading aloud and then discussing the passage became an established ritual. Nietzsche proved witty and clever and was a pacesetter in the conversations. He resolved to write five short passages of his own every day and laid a slate next to his bed to note down the ideas that came to him during sleepless nights. He liked to play the piano, especially when his friends urged him to do so. Malwida later wrote about this life with her 'three sons'.

> When we are reunited this way in the evening, Nietzsche sitting comfortably in the armchair behind his eyeglasses, Dr. Rée, our beneficent reader, at the table where the lamp burns, the young Brenner by the fireplace next to me and helping me peel oranges for dinner, I often say, jokingly: 'We truly represent an ideal family; four people who, previously, hardly knew each other, who have no relational bond, no common memories, and who now lead a life together in perfect harmony, in untroubled freedom, satisfying both intellectually and with respect to personal comfort.' (D'Iorio, *Journey*, p. 40)

Although Nietzsche undoubtedly enjoyed the atmosphere and the friendship of the other people in the house, he developed a very different way of behaving. He increasingly used the freedom they allowed each other to head out on his own, preferably early in the morning or late in the afternoon. He would search for rocky coastlines and little beaches, where he would bathe in the sunrise or in the evening light reflected by the sea and the rocks. Or he might spend time in the gardens, beneath olive-laden branches or amid the orange and lemon trees. The passages of writing he brought back are different from his other work. Nietzsche was distancing himself from Wagner's world and imagining for the first time what free thinking might involve. D'Iorio believes Nietzsche's philosophical concept of the free spirit was born on the coast of Sorrento. The endless sea, the play of the waves eternally coming and going – these were the images he relived, and they were interwoven with ideas about freedom as first of all a bodily experience. Metaphors of sea and coast, of ships leaving harbour, and dreams of blissful islands and of volcanos acquired an important place in his thinking. They seemed to arise in him spontaneously whenever he attempted to fathom what free thinking is, what happens to us when we really do think freely, when we truly leave the harbour behind. This led on to ideas about the most diverse subjects, which Nietzsche noted down in one of the notebooks he always had with him. When he got back he would copy them into an exercise book, or onto loose sheets of paper. Just how special these experiences were for him is clear from a letter to his friend Franz Overbeck, written on 28 August 1877 after he returned to Basel and resumed his teaching and research.

> If only I had a little house somewhere. Then I'd be able to walk for six to eight hours every day as I do here, composing thoughts that afterwards, swiftly and with perfect certainty, I would jot down on paper. I did this in Sorrento, I do this here, and as a result I've gained a great deal from a thoroughly unpleasant and gloomy year. (BVN-1877, 654)

That same day he wrote to Erwin Rohde,

> Should I tell you about myself? How I am always on the road, two hours before the sun comes over the mountains, and especially in the long shadows of the afternoon and the evening? How I have thought about many things and feel so rich in myself, now that this year has at last allowed me to lift away the old moss of daily *compulsion* to teach and think? Given the way in which I am living here, I can endure it even with all the pain – which has of course followed me even into these altitudes – but between times there are so many happy exaltations of thought and feeling.[3] (BVN-1877, 656, italics in original)

In a letter to his sister Elisabeth, again dated 28 August 1877, he writes of his return to teaching, 'And now everything should be covered by a blanket of moss again! Utterly repugnant!' (BVN-1877, 657).

Nietzsche linked metaphors of the coast and the sea with three main themes. The first, as we have seen, concerns having the courage to live freely and think freely. He describes freedom of thought as like daring to leave the mainland, or as he puts it,

> We have left the land and have embarked. We have burned our bridges behind us – indeed, we have gone farther and destroyed the land behind us. Now, little ship, look out! Beside you is the ocean: to be sure, it does not always roar, and at times it lies spread out like silk and gold and reveries of graciousness. But hours will come when you will realize that it is infinite and that there is nothing more awesome than infinity. Oh, the poor bird that felt free and now strikes the walls of this cage! Woe, when you feel homesick for the land as if it had offered more *freedom* – and there is no longer any 'land'. (GS, III, §124, italics in original)

The 'land behind us' represents our desire for firm ground. People look for something solid to hold onto in their lives. They adhere to ideas about life that give them a sense of security and of faith in themselves. Above all they become attached to any authority that supports them in this, to religion for example, or science. Nietzsche opposes the security that Christianity claims to offer. The price people pay for it is too high, he says. Recognizing god as a firm footing means believing in a hold-me-tight world, a world behind this world, a heaven that is supposed to be the real goal of life on earth, with god as the symbol and guarantor of our solid ground. Nietzsche realizes that leaving this land behind means embarking on an unprecedented adventure, one that entails great risks. Anyone who dares to set out on such an adventure without a compass or a map will pass through fear and uncertainty. They will struggle to fight off homesickness for solid ground. If we want to go in search of life we need to have the courage to let go not just of god and Christianity but of other supposed certainties that pin us down at the cost of our lives, including science when it presents itself in the guise of absolute truth and knowledge of reality. Or Plato's philosophy, which claims to have discovered, behind the transience and corruption we experience in everyday life, changeless ideas that offer a firm foothold and place the wave-like motion of our days in a very different light. The metaphor of leaving the land behind occurs regularly in Nietzsche's books and in his posthumously published writings. It is as if he continually needed to give himself fresh courage to go out onto the open sea and – like the old man in Hemingway's story – was fearful of coming back empty-handed.

Nietzsche connects other feelings with the open sea as well, feelings of enthusiasm and astonishment, for instance. How marvellous it is to be able to see and experience everything with fresh eyes, to be able to breathe freedom.

> Indeed, we philosophers and 'free spirits' feel, when we hear the news that 'the old god is dead,' as if a new dawn shone on us; our heart overflows with gratitude, amazement, premonitions, expectation. At long last the horizon appears free to us again, even if it should not be bright; at long last our ships may venture out again, venture out to face any danger; all the daring of the lover of knowledge is permitted again; the sea, *our* sea, lies open again; perhaps there has never yet been such an 'open sea'. (GS, §343, italics in original)

Gratitude, amazement, premonitions, expectation – these are feelings that accompany a person for whom everyday life has become once again an encounter with the great unknown, an adventure. Nietzsche conceives of the 'free spirit' as an outsider who experiences the world through the senses, who becomes sensitive to colours and smells, to the subtle play of change, whether in temperature, in the angle of the light, or in the seasons. Free spirits are utterly unlike armchair scholars. They embody themselves, starting to think and observe and experience in a more bodily manner. They are wanderers, walkers in the outdoors who continually seek contact with an immeasurable earthly reality greater than a human being can deal with or comprehend.

As well as daring to let go and living and thinking more physically, walking along the coast unlocked a third insight: the silence of nature is more eloquent than human speech. This is beautifully expressed in the aphorism that opens the final part of Nietzsche's book *Daybreak*, headed 'In the great silence'. It is a long passage, but so rich that I quote it here almost in its entirety.

> Here is the sea, here we can forget the city. . . . Now all is still! The sea lies there pale and glittering, it cannot speak. The sky plays its everlasting silent evening game with red and yellow and green, it cannot speak. The little cliffs and ribbons of rock that run down into the sea as if to find the place where it is most solitary, none of them can speak. This tremendous muteness which suddenly overcomes us is lovely and dreadful, the heart swells at it. – Oh, the hypocrisy of this silent beauty! How well it could speak, and how evilly too, if it wished! Its tied tongue and its expression of sorrowing happiness is a deception: it wants to mock at your sympathy! – So be it! I am not ashamed of being mocked by such powers. But I pity you, nature, that you have to be silent, even though it is only your malice which ties your tongue; yes, I pity you on account of your malice! (D, §423)

Elsewhere Nietzsche speaks of the earth as the 'great silence' or 'great stillness' that does not speak, cannot speak. All truths that humans think they have discovered about the earth up to now are at best penultimate truths, never ultimate truths. By stripping the earth of its human trappings, Nietzsche enables our planet to appear as the awe-inspiring, the other, the greater, as a reality that does not allow itself to be appropriated or possessed. In his notebooks of the time (1881) he repeatedly writes of his new plan of work: 'My task: the dehumanization of nature and after that the naturalization of humanity, after it has made the pure concept of "nature" its own' (NF-1881, 11[211]).

In encounters with the 'great silence' he becomes increasingly enthusiastic, because the earth proves a formidable opponent, even for him. Speaking against nature immediately rebounds on Nietzsche himself. After all, if we humans inevitably 'humanize' nature and capture it in human speech, then just how valuable are Nietzsche's own words and images concerning nature?

Daybreak continues as follows:

> Ah, it is growing yet more still, my heart swells again: it is startled by a new truth: *it too cannot speak*, it too mocks when the mouth calls something into this beauty, it too enjoys its sweet silent malice. I begin to hate speech, to hate even thinking; for do I not hear behind every word the laughter of error, of imagination, of the spirit of delusion? Must I not mock at my pity? Mock at my mockery? – O sea, O evening! You are evil instructors! You teach man to cease to be man! Shall he surrender to you? Shall he become as you now are, pale, glittering, mute, tremendous, reposing above himself? Exalted above himself? (D, §423, italics in original)

Nietzsche realizes better than anyone that every word, every image falls short; they are merely disguises, falsifications, in fact speaking and thinking are defective modes that can stand in the way of access to nature. Henceforth this makes even thinking itself, or at least what we generally understand as thinking, hard for him to stomach. Nietzsche deliberately goes in search of a more personal, autobiographical and at the same time more expressive language, Colli writes – a language without abstract truth concepts.

Nietzsche in the mountains

One striking thing about Nietzsche's first stay at the coast is that animals seem to have been almost entirely absent from it. He writes of the experiences of freedom and wanting to be free, of space and openness that present themselves as the first portents of contact with the earth, awakening a desire for a different life and different thinking.

Nietzsche was to act upon this desire with a radical move when, three years after his first visit to Sorrento, in the summer of 1879, he left Basel for good. He first travelled to the famous Alpine resort of Saint Moritz in Switzerland but in the end he found it too busy. Wanting an even quieter resort he discovered Sils Maria, a small village not far away in the Upper Engadin, a place with a high, wide view of the mountains and the silent gleam of the Silvaplana mountain lake. Nietzsche immediately recognized these surroundings as the place on earth that he had been seeking. On 23 June 1879 he wrote to Overbeck, 'But now I have taken possession of the Engadin, and am as though in my element, quite marvellously! I am related to this kind of nature. Now I sense an alleviation. Oh, how I have longed for it!'[4] (BVN-1879, 859) Nietzsche returned there for several months every summer, with the exception of 1882. In letters to his friend and assistant Heinrich Köselitz, he spoke of Sils Maria on 8 July 1881 as 'the sweetest corner of the earth' and on 1 July 1883 as 'the place where eventually I wish to die' (BVN-1881, 122 and BVN-1883, 428). His letters are full of intense experiences of happiness, as if his inner landscape were making contact with the landscape around him. To another friend, Carl von Gersdorff, he wrote in late June 1883, 'Oh, how everything is still hidden within me, wanting to become word and form! It cannot be silent and high and solitary enough around me, so that I can understand my deepest inner voices' (BVN-1883, 427).

In Sils Maria Nietzsche stayed in Haus Durisch, where he always rented the same small, dark, unheated room at the back of the house, right up against the mountain. Paul Deussen, a student friend, visited Nietzsche there in 1887 and afterwards wrote,

> It was a simple room in a farmhouse, three minutes from the road: Nietzsche rented it during the season for one franc per day. The décor was as simple as one could imagine. On the side stood his books which I mostly still remembered from earlier times; then came a peasant-style table with a coffee cup, eggshells, manuscripts and toilet requisites in a colorful jumble extending to the unmade bed, via a boot j-jack with a boot jammed on it.[5]

Deussen, a professor himself by this point, regarded the room as evidence of the deplorable situation in which his one-time friend had ended up, but the fact that Nietzsche repeatedly returned to it whereas, as we have seen, he could not find peace at the coast, surely indicates that the discomforts were less important than the atmosphere and the silence offered to him by this little boarding house and its surroundings. He conceived his most important books of the period there and even wrote some of them in full: *The Gay Science, Thus Spoke Zarathustra, Beyond Good and Evil, The Genealogy of Morals, The Case of Wagner, Twilight of the Idols, Dionysian Dithyrambs* and *The Antichrist*.

What did the stay in the mountains and on the shores of Lake Silvaplana mean to Nietzsche? I will group my initial impressions around four points. First, Nietzsche describes repeatedly how the immediate experience of being outside in the mountains vitalizes him and helps him to live. The daily walks, the rhythm of walking, the climbs and descents, the continual succession of new insights that walking in the mountains conjures up, the changes of temperature, of sun and shade, warmth and cold, feeling the fresh air of the mountains that he comes to associate with philosophical thought: it is these fundamental impressions that Nietzsche seeks out time and again and that merge with his observations, feelings, associations and ideas. His interior life increasingly resonates with impressions from outside. Walking slows his thoughts and intensifies his reflections. It even comes to determine the rhythm of his thinking, or what he calls thinking. In his lectures, Dutch philosopher Paul van Tongeren observes that Nietzsche's writing is really best read aloud. You can still hear in it the rhythm of walking. For me, too, these experiences are more than merely context. They form an essential layer of humus in which he seems to make contact with the earth. In those moments of contact, intuitions are born that become the guiding principles of his new vision of the earth and of an 'earthly' life. 'It cannot be quiet and high and solitary enough around me' (1883 letter to Gersdorff, BVN-1883, 427). Being in nature in this way allows his inner voice to be heard. Using more abstract terms, one might say that in Nietzsche the climatological, the spatial, the rhythmical and the musicality of silence become interwoven. The coming together of these experiences awakens powers in him that he refers to as life.

A second striking feature of the mountains is that they bring Nietzsche into contact with animals, which had little if any part to play at the coast. When in his later work he mentions animals in the context of the sea, then apart from birds they are mostly creatures he cannot actually have seen there. He writes of 'the sea, with its rippling snake-skin and beast-of-prey beauty' (HA, §49), which remains alien to man. As far as animals are concerned, the contrast between the mountains and the coast could not be greater.

Thirdly, Nietzsche connects his sojourn in the mountains with a specific spiritual tradition, in which people who want to break loose from ordinary, everyday life traditionally go in search of a different, deeper, higher experience of what being human could mean. This occurs in a number of spiritual traditions, some from the Middle East, such as Christianity and the religions of Persia, others from the Far East, including India and China. Along with the desert, the mountains are the place where the searching human, soberly and without creature comforts, goes to live, choosing the life of a hermit, often accompanied by animals, afterwards returning to the company of humankind transformed. We have already seen that the book *Thus Spoke Zarathustra* belongs to this genre. It is a poetic and prophetic mountain book. Nietzsche avails himself of the classic spiritual framework of inversion and transformation in a way all his own, to give expression to the change he has undergone. I would venture to say that this symbolic–spiritual language comes closest to what Nietzsche experienced as his own most profound change. Moreover, he expects that this genre, this style of writing – although it is more than merely a style of writing – will be the best way to convey to his readers the radical nature and profundity of the metamorphosis he has in mind. We know that in the period in which he composed *Thus Spoke Zarathustra*, Nietzsche was immersing himself in the scientific literature about biology, the theory of evolution and the like, but it is striking that the language of science and the method of arguing and criticizing that goes with it are absent from the book. At best we occasionally hear an echo of it in the distance, for example when modern humans are presented as 'a rope fastened between animal and overman'[6] (Z, I, 'Zarathustra's Prologue, §4').

In *Thus Spoke Zarathustra*, Nietzsche's main work about the new philosophy of the earth, there is no engagement with the content of the scientific literature he was exploring. The book, as I have said, marks a change of style in Nietzsche's writing. Readers can see this for themselves and all the experts agree. Wilfred Oranje, responsible for a new Dutch translation, writes in his afterword that it floats between Nietzsche's other works 'like an ice floe' (*Aldus sprak Zarathoestra*, p. 325). But I believe this is insufficient to explain what is so different about it, even if we add that the style is poetic or metaphorical. Those formal features are deployed at the service of something that is entirely new both in its content and in a philosophical sense, an attempt to put into words the profundity of an experience and the task with which it presented him. Nietzsche is looking for a spirituality that takes the bodily aspect of life as its guide, rather than as material to be processed into something higher.

This brings me to a fourth and final element of Nietzsche's journey into the mountains. He said and wrote that he had an extraordinary and intense experience during his first stay in Sils Maria. Something exceptional happened to him there. It took place during a walk along the shores of the mountain lake, and Nietzsche carefully recorded it in every detail. The date was 6 August 1881. On his way from Sils Maria to Silvaplana–Surlej he stopped at a pyramid-shaped rock. There he was suddenly struck, as if by lighting, by an insight that he summed up in the words '*Ewige Widerkehr des Gleichen*', usually translated into English as 'eternal recurrence'. As his biographer Rudiger Safranski writes, 'It was immediately apparent that his life was now divided into two halves: before and after the inspiration.'[7] In Chapter 3 I return to this experience, but now I want to point out two features that are of relevance here. First of all, it was an experience in nature and of nature. During a walk high in the mountains, this insight – which is more than merely an insight – struck home and turned Nietzsche's life upside down. Secondly, this intense experience, which permeated his entire body, inspired him to write *Thus Spoke Zarathustra*. Two years later, on 3 September 1883, he wrote to Köselitz, 'This Engadin is the birthplace of my Zarathustra. I recently found the first draft of the thoughts that are collected in him; underneath it reads "beginning of August 1881 in Sils Maria, 6,000 feet above the ocean and far higher still above all human things"' (BVN-1883, 461). In other notes too, Nietzsche gives the same date and place. The fact that he continually refers back to this event with such specificity makes his description consistent with other accounts of mystical experiences. Others too write that the moment is engraved on the body, marking a turning point. Whatever our attitude to such testimony, it is reasonable to conclude that Nietzsche had a profound experience of the natural world.

Both during his stay in the mountains and in his time on the Mediterranean coast, Nietzsche soaked up nature, physically and mentally. He came to see himself as a participant in the lives of animals and plants, of the mountains and the sea. This way of life was the first step towards his 'naturalization', as he called it. The intense experiences he encountered during his walks, his sharp observations and his meditative daydreams all laid the foundations for the way in which Nietzsche looked at life from then on. His philosophizing became interspersed with comparisons derived from nature, and animals commonly served as examples: 'like the lion . . .', 'like the eagle . . .'. The model of comparison, of analogy and metaphor, became a step towards philosophical reflections about Nietzsche's own life and the lives of the people around him. But this entire transformation was embedded in an even larger project, a quest to free himself from the illnesses, both physical and cultural, that were preventing him from living. The aim of his whole undertaking was 'the great health', and the way to it is pithily described in an aphorism in *Daybreak* that he entitles 'By circuitous paths':

> Whither does this whole philosophy, with all its circuitous paths, want to go? Does it do more than translate as it were into reason a strong and constant drive, a drive for gentle sunlight, bright and buoyant air, southerly vegetation, the breath of the sea, fleeting meals of flesh, fruit and eggs, hot water to drink, daylong silent wanderings, little talking, infrequent and cautious reading, dwelling alone, clean, simple and almost soldierly habits, in short for all those things which taste best and are most endurable precisely to me? A philosophy which is at bottom the instinct for a personal diet? An instinct which seeks my own air, my own heights, my own kind of health and weather, by the circuitous path of my head? There are many other, and certainly many much loftier sublimities of philosophy, and not only those which are gloomier and make more claims for themselves than mine – perhaps they too are one and all nothing other than the intellectual circuitous paths of similar personal drives? In the meantime I have come to look with new eyes on the secret and solitary fluttering of a butterfly high on the rocky seacoast where many fine plants are growing: it flies about unconcerned that it has but *one* day more to live and that the night will be too cold for its winged fragility. For it too a philosophy could no doubt be found: though it would no doubt not be mine. (D, §553, italics in original)

It is a beautiful passage, but one that leaves us with a question. Is this coupling of life and thought tailor-made for Nietzsche and suitable only for him? Or does what he writes contain a call for a new way of philosophizing? In the chapter that follows it will become clear that Nietzsche did not limit himself to emotion, however deep and rich his feelings might be. Epicurean enjoyment and a sojourn in the garden of nature were not enough for him. He was gripped by the need for us to change both our lifestyle and our culture. His ambitions reach even further. He went on to ask whether the relationship between inside and out, between how people live and what surrounds them, can be influenced, whether in a positive or in a negative sense.

2

Interacting with the natural world around us
On personal lifestyle and local culture

From the 1880s onwards, Nietzsche became fascinated by all kinds of metabolic processes that occur between humans and their environment. He was interested in the way in which people engage with the earth and make it their own. On this subject too he tried to stay as close as possible to life, to everyday experience. He was intrigued, for instance, by the way people create gardens and by doing so make a stretch of uncultivated land part of their lives. He saw the garden as a space where humans and the earth develop a mutual connection. The garden was both a microcosm, all of nature in miniature, and an autonomous creation by the hand of man. Yet it was not gardens that fascinated him most but nutrition, the way people digest the fruits of the earth, thereby making them part of their bodies. When enumerating the issues that had been most important in his life, Nietzsche often put this in first place. The salvation of humankind depended upon nourishment more than anything else, which naturally includes what we eat and drink and how food is prepared, but also the way we cook, the times at which we eat, the regularity and rhythm of eating. Does the food that becomes part of us affect how we feel, how we speak, think and fantasize? This whole complex subject became a personal field of research for Nietzsche. He drew a distinction between *Körper* and *Leib*, between the body as an object and the lived and living body. The first of these two nouns belongs to the language of medicine, or biology, or so Nietzsche claimed. The word '*Leib*' is more primary, closer to our experience – at least that is how we need to understand it if we are to comprehend Nietzsche's determination, frequently expressed in his notes, to develop a philosophy that adhered to the 'guide of the body' (*Leitfaden des Leibes*). Food and the garden became the subjects that for him symbolized more than any others the interaction between humans and their environment.

To get a clear view of this interplay, Nietzsche introduced two new metaphors. The first is '*einverleiben*'. Since the English language does not precisely mirror the words '*Leib*' and '*Körper*', it is tempting to preserve Nietzsche's distinction by translating the verb *einverleiben* as 'to assimilate', meaning to take up into the living body, including the life of the mind and spirit, thereby leaving 'to incorporate' to refer to absorption into the body-object. That, however, would be to lose sight of the importance of the bodily aspect in Nietzsche's thinking, so 'to incorporate' will be used in this book as a translation of *einverleiben* and the reader relied upon to recall Nietzsche's resistance to naive Cartesian dualism. For Nietzsche, the living body could not be separated from the mind. It is this incorporation that Nietzsche decides not only to observe more precisely but to subject to what he calls experimentation.[1]

The other metaphor Nietzsche uses in thinking through human interaction with the environment is gardening. He uses it to illustrate how people deal with nature, enjoy it and feel absorbed by it, as well as how they continually engage in battle with the elements and the forces of nature, pitting their strength against them and displaying their inventiveness. Gardening proves the perfect metaphor for human creativity.[2]

How important this new focus on the body and on experimentation with food and the soil was to become for Nietzsche is usefully explained by philosopher Giorgio Colli in his book *Scritti su Nietzsche* (1980), translated into French as *Écrits sur Nietzsche* (1996). He writes that in the 1880s Nietzsche became rather annoyed by what he experienced as the tossing back and forth of theoretical ideas and abstract concepts. He had tried all possible combinations and sometimes experienced his thinking as a game with empty symbols. Full of his own words, friendless, ploughing an increasingly lonely furrow, Nietzsche turned his attention to his personal way of life. 'With this posthumous material . . . we have the impression that we are entering the laboratory where he experiments with new techniques. . . . Nietzsche goes inside himself and discovers new landscapes,' Colli writes (1996, pp. 150 and 152). The laboratory to which Colli refers is Nietzsche himself, life and limb, including his lifestyle and habits.

Why experiment? Because Nietzsche's search was always led by the following questions: Is there a better way? Can we improve our methods of nourishing ourselves and of gardening to such an extent that nature becomes more intensely and consciously part of our lives? Can we invent a lifestyle that truly brings us 'great health'? Nietzsche goes in search of an idea of becoming human that he characterizes as an attentive, creative, caring and inventive incorporation of nature into his life and the interweaving of his life with nature. In place of the word 'nature' I could have used the word 'earth', since Nietzsche uses both terms when he writes about the process of incorporation.[3] This second aspect, becoming part of the greater process of life on earth, belongs with the first, as we shall see. Already it was becoming clear that he wanted to understand becoming human as a dual process of *einverleiben* or incorporation: of the earth in the human being and of the human being in the body of the living earth.

Nietzsche puts it like this:

> *In media vita.* – No, life has not disappointed me. On the contrary, I find it truer, more desirable and mysterious every year – ever since the day when the great liberator came to me: the idea that life could be an experiment of the seeker for knowledge – and not a duty, not a calamity, not trickery. – And knowledge itself: let it be something else for others; for example, a bed to rest on, or the way to such a bed, or a diversion, or a form of leisure – for me it is a world of dangers and victories in which heroic feelings, too, find places to dance and play. '*Life as a means to knowledge*' – with this principle in one's heart one can live not only boldly but even gaily, and laugh gaily, too. And who knows how to laugh anyway and live well if he does not first know a good deal about war and victory?[4] (GS, IV, §324, italics in original)

Anyone who experiments must dare to take risks and be aware that even the failure of an experiment can be a result. If the experiment succeeds, who would not have a sense of victory? 'What doesn't kill me makes me stronger,' Nietzsche later wrote.

I want to explore three aspects of this: eating and fasting; the garden as a link between human beings and nature; and finally the climate and the cosmos, which have a special influence on all interactions between humans and the environment. In this chapter I examine how Nietzsche imagines this interaction and how his personal experiments with food and the environment were geared to improving the relationship. The insights he gained became guidelines for his assessments of his own life and of local cultures, based on which he produced advice for the future.

Eating, fasting and diet

Because of his weak constitution and his always intense reactions to everything with which he came into contact, Nietzsche became oversensitive to food. It seemed to influence him more than anything else. Based on earlier experiences of sickness and misery, he composed his meals increasingly carefully and observed with great precision what his body could and could not incorporate. Once he had passed the age of thirty he was forced to conclude that he had little tolerance for even the best alcohol. With barely concealed envy he described how those with stronger constitutions could devour the heaviest, fattiest of food without their bodies protesting. They could digest it, too, he adds – no doubt referring to the German cuisine of his childhood. These personal circumstances helped to fuel Nietzsche's fascination, as a philosopher, with the entire process: the selection and preparation of food, eating, the absorption of food by the stomach and intestines, and the excretion of that which has not been digested, assimilated and transformed into part of the body. That philosophers paid so little attention to this essential function surprised him at an early stage. In *The Gay Science* he writes,

> So far, all that has given color to existence still lacks a history. Where could you find a history of love, of avarice, of envy, of conscience, of pious respect for tradition, or of cruelty? Even a comparative history of law or at least of punishment is so far lacking completely. Has anyone made a study of different ways of dividing up the day or of the consequences of a regular schedule of work, festivals, and rest? What is known of the moral effects of different foods? Is there any philosophy of nutrition? (The constant revival of noisy agitation for and against vegetarianism proves that there is no such philosophy.)' (GS, I, §7)

In the 1880s this general question became focused on Nietzsche's own lifestyle. He started consciously experimenting with his diet, with more or less salt in his food, more or less spicy dishes, meat or the avoidance of meat. This experimentation continued for a long time, and in *Ecce Homo*, one of his last and most autobiographical books, Nietzsche looks back at the role food has played in his life. He mentions it first in a summary of the issues that have been most important to him. 'One can for convenience' sake formulate it thus: "how to nourish yourself so as to attain your maximum of strength"' (EH, 'Why I Am So Clever, §1'). For Nietzsche, this ultimately became a key question in relation to food, so he makes the issue a personal one, addressing the notion of vitality. He gives a detailed account. 'My experiences here are as bad as they possibly could be; I am astonished that I heard this question so late, that I learned "reason" from these experiences so late.' He lays the blame first of all at the door of German cuisine.

> German cookery in general – what does it not have on its conscience! Soup *before* the meal (in Venetian cookery books of the sixteenth century still called *alla tedesca*); meat cooked to shreds, greasy and floury vegetables; the degeneration of puddings to paperweights! If one adds to this the downright bestial dinner-drinking habits of the ancient and by no means only the *ancient* Germans one will also understand the origin of the *German spirit* – disturbed intestines. (EH, 'Why I Am So Clever, §1', italics in original)

He is barely any kinder about English or French cuisine, incidentally. The best cookery as far as he is concerned is that of Piedmont in northern Italy. As regards alcohol, Nietzsche naturally drank a good deal as a student and discovered that too strong a dose virtually transformed him 'into a sailor'. But later, 'towards the middle of life', he turned against alcohol. He writes that he cannot advise too strongly all 'more spiritual natures' to 'abstain from alcohol absolutely'. 'Water suffices . . . I prefer places in which there is everywhere opportunity to drink from flowing fountains' (EH, 'Why I Am So Clever, §1'). He was referring to Nice, Turin and Sils. He ends this section on food with several more tips, including the following: know the size of your stomach, do not eat between meals, avoid coffee, and drink tea only in the mornings and 'not the slightest bit too weak'.

Why is food so important to Nietzsche that he mentions it first? Because it is the most substantial and all-pervading link between humans and their place on earth. What we eat is transformed to become part of the body. Nietzsche describes this process in detail: how the stomach is able to assimilate the most varied foods or otherwise reject them, and how the intestines unconsciously do their dark, impenetrable work, 'digestion' invisibly becoming part incorporation and part excretion. Ingestion, assimilation, incorporation and excretion are the core activities of the process of nourishment in which life reveals itself as an interplay of forces and impulses that transform the products of the earth into energy and so 'ground' our bodies. Upon this insight, which has since become fairly obvious to us, Nietzsche builds his argument for understanding the various cultures as first of all ways of feeding ourselves.

Hence his attention to the characteristics of local cuisines, be they German, French or Chinese, and to the fact that geographical and regional differences in diet can make us conscious of the connections between food, region, physicality and the art of living well.

The second reason why diet is so important to Nietzsche concerns the fact that the quality of our food, vital to our very existence, is subject to human influence. Nietzsche searches for examples and ideas as to what a conscious culture of nourishment and eating might look like. He advocates a health-giving nutritional environment, taking account of the fact that from time to time it is good not to eat. 'I also want to make asceticism natural again: in place of the aim of denial, the aim of strengthening; a gymnastics of the will; abstinence and periods of fasting of all kinds, in the most spiritual realm too; a casuistry of deeds in regard to the opinions we have regarding our strengths; an experiment with adventures and arbitrary dangers'[5] (NF-1887, 9[93]).

Nietzsche describes not just material food in terms of assimilation and incorporation but 'mental' food as well, in other words the information the media serve up to us daily, and ideas about all kinds of things that 'enter' hour after hour, along with the impressions and emotions that come with them and penetrate our bodies. Modern culture was hyperactive on this point, Nietzsche believed, even in the late nineteenth century. The number of impressions was increasing and the pace at which they were presented and needed to be absorbed was accelerating by the day, overfeeding people. 'The overabundance of diverse impressions is greater than ever: cosmopolitan in food, literature, newspapers, shapes, tastes, even landscapes and so on. The tempo of this stream is prestissimo' (NF-1887, 10[18]; see also NF-1887, 9[165]). People wasted their energy adjusting to this overabundance of impressions. They forbad themselves to truly take things in, to digest them, and this weakened their ability to digest anything at all. They became oversensitive without even realizing it. Mental food too is part of the metabolic process that occurs between humans and their environment. Here Nietzsche advises us to stop swallowing everything unconsciously and therefore indiscriminately. A new culture requires a more alert and deliberate way of dealing with both material and mental nutritional products. We will have to learn to monitor the stream of impressions we allow in and their quality, based on the questions: What makes me stronger? What increases my vitality? What suits me?

It is precisely on this point that we can learn from the animals. Nietzsche calls on the cattle to help us to learn how to ingest food slowly, making clear to us that re-chewing food, or 'chewing the cud' as cows do, whether the food be material or mental, is extremely important and takes time. Whenever he wanted to evoke the new culture that we modern humans will need more than ever in future, a culture in contact with the earth, Nietzsche deployed terms like slow down, reduce, select, ruminate, digest and incorporate.

In sum, the language of nutrition, eating and digesting, of fasting and creating a healthy diet, became the language Nietzsche would turn to when he wanted to contemplate the fundamental metabolic processes that occur between humans and their earthly environment. In increasingly concrete metaphors and stories, he tried to visualize and describe that relationship and support it with his own empirical observations and discoveries, and the results of his experiments. 'I would wish that, if only once, I had met a person who asked of everything that came into his hands: Could this not be improved upon? The meals and the diet, the schedule for the day and so on' (NF-1881, 14[11]). All this comes under the heading of a new culture, a new art of living well, in which a lived and carefully considered food policy in the broadest sense would have to take shape. Because, Nietzsche wrote, 'I live so that I can discover; I want to discover so that the overman can live. We are experimenting for him' (NF-1882, 4[224]).

Gardens and gardening

Another way of bringing the body into contact with the earth and of 'incorporating' the earth is the art of gardening. The garden is a space where humans and the earth influence each other intensively. In the garden, nature and artisanal creativity, joy in life and the desire for order and productivity all come together. From the biblical Garden of Eden to the utopia of Walden Pond, the garden has prompted religious, moral and philosophical reflections on how people ought to treat the natural world and conversely how the natural world enables people to discover their own true natures. Nietzsche places himself in this tradition when he chooses the garden – as a reality and as a metaphor – as the basis from which to develop his vision of an active, experimental interchange between humankind and the earth, between nature and culture.[6]

The theme of the garden first arose in the years after Nietzsche's stay in Italy, taking long walks through the suburbs of Genoa and Turin, with their big houses surrounded by carefully laid out gardens in the Italian style. Nietzsche may well have chosen these walking routes because he was reading *The Cicerone*, a book about Italian art that was much read in intellectual circles at the time. It was written by Jacob Burkhardt, a fellow professor at Basel, and regarded as a masterpiece of writing on the subject. It pays a good deal of attention to the Italian villas and their gardens and examines the various styles of gardening. These included what would later become known as the English style, a type of landscape gardening that was immensely popular in the nineteenth century for its apparent naturalism. A landscape garden had narrow paths winding through it; there were groups of trees, stretches of lawn, shrubs, and often a small artificial lake, or a pond with a waterfall. The intention was to give the impression of being surrounded by nature. It is a style that aims to suggest both naturalness and freedom, and in doing so differs markedly from the traditional French garden, which stresses authority and hierarchy. The gardens of Versailles that surround Louis XIV's palace are a good example of the latter. Burkhardt did not share this fondness for the English landscape garden, which he regarded as kitschy and sentimental, even deceitful, because everyone could see that it was not real nature but a pretence. He preferred the gardens of the Italian Renaissance.

Nietzsche was clearly influenced by Burkhardt's view. He adopted the same perspective, and in the gardens around the villas on the outskirts of Genoa he saw two qualities come together, both of which spoke to him. In contrast to the quasi-naturalism of the English landscape garden, Italian gardens were obviously deliberately laid out. The creative hand of the architect remained visible in the design, not disguised by fakery. Furthermore, Italian gardens were created in such a way that they offered a view of the natural world beyond, the sea and the mountains. They were the epitome of places where people and the earth come into contact and enter into a relationship, continually conscious of each other's presence. We might say that during his walks around Genoa and Turin, Nietzsche became interested in what is now known as landscape architecture. He expressed his admiration for an art of gardening that involved the land in the lived environment in such a way that the city dweller remained open and sensitive to broader connections with the mountains and the sea.

Genoa. – For a long while now I have been looking at this city, at its villas and pleasure gardens and the far-flung periphery of its inhabited heights and slopes. . . . I keep seeing the builders, their eyes resting on everything near and far that they have built, and also on the city, the sea, and the contours of the mountains. (GS, IV, §291)

He looked at houses and gardens and asked himself: What values are being expressed here? The insightful information presented by Burkhardt's charting of sixteenth- and seventeenth-century history prompted Nietzsche to document gardening more broadly. He investigated how patches of earth were absorbed into the lived environment in antiquity and by other cultures, and how people have transformed land into gardens, thereby incorporating it into their lives. His interest then extended to larger gardens that project a grand style, such as those of the Roman period or the immense endeavours by the ancient Egyptians. In Nietzsche's writing we find little information about still older forms of human civilization, such as the first agrarian cultures, and their development. Even local farming life in his own time was fairly remote from him.

This is one of the subjects that Nietzsche studied for reasons other than an interest in the material as such. He had no desire to become a landscape architect or a historian of gardening. His fascination for the phenomenon of the garden was part of a larger perspective, that of the earth as a whole, and part of a search for a new culture that remained faithful to the earth. The garden became the point of crystallization, the topos around which Nietzsche thought experimentally about an active relationship between human beings and the earth. He related it to his knowledge of various local cultures. But he did all this with one question in mind: How can we learn to live and to cultivate the land around us better? Does a comparison between cultures help us to do so? Zarathustra, Nietzsche's alter ego, would later regale readers at length about the way that the earth not only gives us food but produces it in the form of delightful fruits, with scents and colours that vitalize our bodies, delight our eyes and warm our hearts.

Just how important the garden was to Nietzsche is clear from the fact that he uses the metaphor of gardening when writing about the attitude of human beings to their own natures, their own natural aptitudes and urges. Here too he advocates an active and experimental approach.

One can dispose of one's drives like a gardener and, though few know it, cultivate the shoots of anger, pity, curiosity, vanity as productively and profitably as a beautiful fruit tree on a trellis; one can do it with the good or bad taste of a gardener and, as it were, in the French or English or Dutch or Chinese fashion; one can also let nature rule and only attend to a little embellishment and tidying-up here and there; one can, finally, without paying any attention to them at all, let the plants grow up and fight their fight out among themselves – indeed, one can take delight in such a wilderness, and desire precisely this delight, though it gives one some trouble, too. All this we are at liberty to do: but how many know we are at liberty to do it? (D, §560)

In this aphorism, Nietzsche makes clear that humans, in philosophy especially, are not used to imagining human nature as an expanse of land ripe for cultivation. In contrast to thinkers who conceive of human nature as a pre-defined pattern or believe that a person simply cannot be changed, Nietzsche attributes to human nature uncultivated natural powers and believes every person has the freedom to shape them. How we do so is up to us. It is for the gardener to determine what design or what local customs to take as a source of inspiration, or indeed whether to undertake anything at all.

By characterizing human nature as an expanse of uncultivated land, Nietzsche also makes clear that the body has every characteristic shown by the earth and is composed of the same materials as everything else on the planet. 'How strange and superior we are towards the dead, the inorganic, and meanwhile we are three quarters a column of water, and have inorganic salts in us that probably have more influence on our fortunes than the whole of living society' (NF-1881, 11[207]). A little later he adds, 'The inorganic determines us through and through: water, air, earth, the composition of the soil, electricity and so on. We are plants under such conditions' (NF-1881, 11[210]). But the inorganic comes to life in humans in a complex interaction between urges. Nietzsche sees the human being above all as an organic creature, as a complex, continually changing interplay of passions, an entity made up of bodily potentials, not based on any universal model. Unlike the many thinkers who believe human nature to be uniform, whether in an Aristotelian or Kantian or more positivist or genetic sense, Nietzsche defends a completely open idea of the nature of every person, even though each of us carries the genetic traces of the generations that produced us. He qualifies this in just one respect: in contrast to all other organic beings, every human complex of passions needs to take shape, one way or another. Typical of humans is that they have to be moulded. It may be that their form is imposed from outside the individual, by upbringing and culture, but – and of course this is Nietzsche's main concern – those who are strong enough can learn to shape themselves, their passions and their lives, which is where the gardening metaphor comes in. The interplay of the passions does not grow into something good and beautiful by itself. We are in trouble if we are only the soil of our garden and not also the gardener, Nietzsche writes in other aphorisms. For example: 'Gardener and garden. – Out of damp and gloomy days, out of solitude, out of loveless words directed at us, conclusions grow up in us like fungus: one morning they are there, we know not how, and they gaze upon us, morose and grey. Woe to the thinker who is not the gardener but only the soil of the plants that grow in him!' (D, §382)

It is striking that whenever Nietzsche writes about human nature he mentions concrete experiences or emotions. They seem chosen almost at random, as if to prevent us from associating human nature with any distinguishing features. This randomness also suggests we must think at that same everyday level. What do I grow if I feel gloomy, or lonely or cold-hearted? What thoughts and feelings shoot out of the ground like mushrooms? What do they reveal about me as a gardener? How do I process them? A few years later, Nietzsche used the dual character of the gardener and the garden in a broader metaphor that afterwards returns several times in his writing: 'In man there is also creator, sculptor, the hardness of the hammer, the divine spectator and the seventh day – do you understand this antithesis?' (BGE, §225).

Humans are a combination of two completely different sets of components. On the one hand there are things that have the lowest possible status in philosophy, such as clay, mud, dust, urges, chaos, including the chaos of thoughts and feelings, and unordered excess. On the other there are things regarded as the most elevated of forces: creativity, art, the wielding of the hammer, the superior attitude of the observer and resting on the seventh day. Nietzsche brings the two together and combines them by means of the radical freedom to experiment, because that is the approach that fits best with what we know so far and above all what we do not yet know about nature. This framework opens up Nietzsche's own perspective, his own vision of the art of gardening. He does not attempt to attribute a new value to wilderness, as Thoreau did in America, as if it were nature in its purest manifestation.

Even less does he imagine nature as a beacon of innocence, an idealization along the lines of the romanticism of Rousseau that regards all attempts to impose form as distortions that originate in society. Nietzsche works out a perspective in which he can hold two opposing poles together that were previously mutually repellent. On one side is unspeaking, open nature, imagined as a complex interplay of forces, and on the other the absolute necessity of humankind to shape, to order, to stylize in freedom. But – and this is Nietzsche's most important addition – this stylization has to be guided by values that must be discovered and invented in a process of dealing with the earth, a process that continually purifies itself. Nietzsche uses the image of the gardener and the garden as a metaphor for the never-ending dialectic between being surprised by what nature allows us to discover and liberating ourselves from all the impedances that stand in the way of such open exchange. Those who dare to set to work in this way, who dare to experiment with themselves and their environment, are working to lay a path for the coming of the overman.[7]

In a book that remains highly informative to this day, *Nietzsche over de Menselijke Natuur* (*Nietzsche on Human Nature*, 1994),[8] Joep Dohmen points to the importance of the 'splendid yet rarely noted' metaphor of the garden in Nietzsche. He describes it as the theoretical basis for Nietzsche's vision of an emancipatory art of living well and shows how Nietzsche uses it first to get nature under control and then to give it shape, making it into his own garden creation. Dohmen quotes passages such as 'We have the capacity to cultivate our temperament like a garden, to plant experiences in it and weed others out, to establish a beautiful quiet avenue of friendship, to become conscious of secluded prospects of fame – maintain access to all these good little nooks in your garden, so that they are available when we need them' (NF-1880, 7[211]). In his final chapter, which he entitles 'Nietzsche's plea for a second nature', Dohmen makes gardening the basic model for the training in refinement to which all humans must have the courage to subject themselves. He writes, 'Following on from these garden metaphors, Nietzsche works out a concept of refinement in two ways: on the one hand in the narrower sense in which the aim is to pay attention to one particular passion; on the other hand in a broader sense, so that the system of passions as such is raised to a higher level' (1994, p. 481). For Dohmen, Nietzsche's art of living well is to be found in the task of all humans to mould themselves, to give their passions and eccentricities, and their hidden talents, a unique form suitable only to themselves. This is certainly correct, but does Dohmen go far enough in his description of the Nietzschean art of living well? Is Nietzsche saying: make your life an artwork, of no matter what kind? Or is this always part of a wider perspective on the mountains and the sea, which is to say the earth? Do we hear through it the question on which everything in Nietzsche turns: What art of living well is at the service of the overman, of the human who is in contact with the earth?

In Nietzsche's work, the metaphor of the garden clearly fulfils a number of functions. What they all have in common is that Nietzsche sees the relationship between humans and nature – nature within themselves and nature around them – as interactive in the literal sense. Both poles, humans and the earth, point to actors, because the earth, in its muteness, is imagined as a partner. The two actors influence one another and as they do so both are changed, in a continual process that takes shape in material and cultural forms of life. Nietzsche is interested in all these forms, and in the background there is always the question: How could this be improved? How could the interaction between humans and the earth progress in such a way that people form themselves into powerful and healthy creatures and allow the earth to emerge with all its hidden potentials? Nietzsche looks forward to a culture that shapes its members into practised 'gardeners' who have learnt to live in such a way that both types of gardening reinforce each other and together create a framework for living on the planet as people who are in contact with the earth.

In 1990 Dutch writer Gerrit Komrij gave a Huizinga Lecture with the title 'On the Necessity of Gardening'.[9] He reconstructed the garden myth from the perspective of art and the artist, and described how fruitful and thrilling the symbolism of the garden can be as it is revived time and again in classical philosophy, in the religious and biblical literature of the Middle Ages, in the Renaissance garden, and then in sixteenth-, seventeenth- and eighteenth-century designs, right through into the romanticism of the nineteenth century. Seen in this way, the history of the garden is a history of changing ideas about contact between humans and nature, he said. But he regretfully concluded that in the twentieth century the myth of the garden had lost its power.

> We treat our parks and the last of our public gardens atrociously. Look at today's strips of green with their horrible council shrubs. See today's gardens in their desperate efforts to express some individuality of their own as puny patches laid side by side: a kind of agitated and agitating macramé. They lack any connection with a higher idea and if they involve any hint of

anything from garden design so ghastly as a 'total concept', then they still look more than anything like annexes to garden centres and DIY shops. (Komrij 1991, p. 61)

What values are expressed by gardens in our own time? Today, can the garden still evoke the image of a connection between our inner being and the cosmos? Komrij went on to say,

Even if we conceive of gardening at the simplest level, as the pleasant feeling of actively collaborating with the world of nature and of planting in it the living nature of humans, we are forced to conclude that the myth of humans as gardeners has become impotent. It was a myth that created not just cohesion but a suggestion of responsibility. (Ibid., p. 63)

Komrij called upon the contemporary artist to become 'a humble gardener' and start in the library, 'to search, dig out and fertilize what has disappeared from sight and been lost'.

Climate and cosmos

Interactions between humans and nature are embedded in dynamic climatic and cosmic processes that people in Nietzsche's day could not have grasped. It is obvious that changes in weather conditions, the succession of seasons and the rhythm of day and night have a major impact on how people live. As the years went by, Nietzsche was increasingly aware of how important these interactions would become. There is every indication that he was extremely sensitive to the effects of climatic and cosmic changes on his experiences, moods and thoughts, and above all on his questions about the relationship between human beings and the earth. Perhaps the greatest revelation that came to him was to feel taken up, incorporated into the larger space of life and connected to the firmament around him. It was there that the healing of his body and mind began. Sun, moon and stars, diurnal rhythms and the order of the seasons became guides in the process of transformation undergone by Zarathustra. The more intensely he went along with the rhythms of nature, the more penetrating became the questions that confronted him. In living with the elements, Zarathustra begins to experience what it means to be an inhabitant of the earth, as we shall see in the next chapter, but already it should be clear that climate and cosmos substantially coloured the perspective from which Nietzsche looked at himself and the cultures around him. The word 'cosmos' refers above all to the heavenly bodies we observe, with which every person and every local community is in direct contact: the sun, the moon and the stars. Through the cosmos, as it were, Nietzsche looked at life on earth.

This cosmic view opens up a perspective that lifts human beings above their personal and local boundaries and connects them with planetary life as a whole. From that perspective, Nietzsche became interested in how cosmic and climatic experiences are expressed in different cultures, in their symbols, their art, their religion and morals. At an early stage he drew a distinction between temperate and tropical cultures, and he was interested in the different symbioses that have arisen between region, morality and climate.

> In comparison with the temperate zone of culture into which it is our task to pass over, that of the past produces, taken as a whole, the impression of a *tropical* climate. Violent antitheses, the abrupt transition of day to night and night to day, heat and vivid colour, reverence for everything sudden, mysterious, terrible, the swiftness with which a storm breaks, everywhere a prodigal overflowing of the cornucopias of nature: and, on the other hand, in our culture a bright yet not radiant sky, a clear, more or less unchanging air, sharp, occasionally cold: thus are the two zones distinguished from one another. (HA, §236, italics in original)

Tropical cultures nourished a very different imaginative realm from those of the moderate cultures of northern Europe, Nietzsche believed. In his later work he tried to devise a typology of cultures with climatological characteristics that were more and more refined, although his efforts were unsuccessful.[10] When in the autobiographical *Ecce Homo* he looked at his own life from this perspective, he suggested that geniuses should ideally make their home in 'Paris, Provence, Florence, Jerusalem, Athens – these names prove something: that genius is conditioned by dry air, clear sky' (EH, 'Why I Am So Clever, §2'). Was this meant to be taken seriously, or is it an ironic way of telling the reader that it was no accident his body recognized a particular place on earth as his own? Patrick Wotling devotes a chapter in his *Nietzsche et le problème de la civilisation* to the climate and concludes that Nietzsche never let go of the metaphor of climatic zones, even though it underwent several changes and became more complex and ambiguous.[11] What does not change is Nietzsche's determination to assess cultures based on the questions: What values are expressed in those particular cultural patterns? How do they contribute to a tried and tested relationship with the earth? He regards precisely that sensitivity as lacking in modern culture. Modern humans have exchanged their old cosmic and planetary consciousness for a lifestyle and culture no longer aware of these things. They have shut themselves up and cut themselves off in cities full of people.

Nietzsche wanted to develop his own planetary perspective, his own 'cosmology' as it is usually called in philosophy. He therefore consulted classical philosophers such as Epicurus and the Stoics, as well as Eastern philosophy. I will look at this in more detail in Part II, but the extent to which his own personal speculation about the climate and the cosmos affected the way he set out on this path of his own can be further illuminated by the following four observations. First, Nietzsche's cosmology is fragmentary. He did not write a systematic treatment of the subject. In his many aphorisms he looks at one new aspect after another, and as a result his cosmology consists largely of a varied collection of separate points, the only exception being the book *Thus Spoke Zarathustra*.

Secondly, Nietzsche sticks as closely as possible to the experiences he gained during his daily walks. He brings the whole range of his personal experience with him and a wide variety of emotions: surprise, astonishment, admiration, fear, melancholy, gratitude and joy. His cosmology starts with these experiences, indeed it is an experiential cosmology. He enables modern humans, by means of experience, to discover how important, how vital it is for them to open themselves up unconditionally to the universe and to feel again for themselves how healing and enriching that can be.

Thirdly, the political context of the time powerfully influenced Nietzsche's thinking about both the always locally situated human cultures and the planetary perspective. In the prevailing mentality of the day, a positive attitude to local traditions had become a political issue. The influence of the state, which increased throughout the nineteenth century, meant that in Germany a reassessment of national traditions and customs became a vehicle for nationalist and even fascist politics. There was growing exaltation at the idea that the Germans were one people and one nation, with a distinctive German identity. Slogans arose about *Blut und Bodem* and *Volk und Vaterland*. Nietzsche radically distanced himself from this tendency. He was sharply opposed, for example, to the glorification of the German Spirit, with which he did not want to be connected in any way. He fiercely attacked this kind of nationalism, partly out of a conviction that people needed to learn to think as Europeans. It was an ability yet to be achieved, a fact that he made clear as follows.

> We 'good Europeans': we too have our hours when we permit ourselves a warm-hearted patriotism, a lapse and regression into old loves and narrownesses – I have just given an example of it – hours of national ebullition, of patriotic palpitations and floods of various outmoded feelings. More ponderous spirits than we may have done with what in our case is confined to a few hours and is then over only after a long period; one takes half a year, another half a life, according to the speed and power with which he digests it and of his 'metabolism'.[12] (BGE, §241)

We also find a very different kind of thinking in Nietzsche about the always local and multifarious character of cultures. We can track it down by investigating how he approached the subject by means of ecological questions and motifs, and by attempting to characterize them with the help of injunctions such as 'remain faithful to the earth'. Both the tone and the content of these passages are entirely different from those of his political writing on the subject of nationalism. So there is in Nietzsche a political–philosophical line of thinking – in which he opposes the tendency to stamp the local with sentiments of appropriation of land and folk tradition and to indulge an aversion to all things coming from outside – and alongside it an ecological line that takes as its theme the relationship between local cultures and the planetary perspective. Both occupied his mind.[13] I limit myself in this book to the second of the two.

Fourthly, Nietzsche tried to develop an open cosmology. In that sense he departed from classical philosophy. To the Stoics, the firmament of sun, moon and stars presented an eternal, unchanging and completed order, of which life on earth was part. They told people that in this order of things the rules could be found that would tell them how to live. Nietzsche took a different idea as his starting point. He believed that we must begin to discover that we cannot have any conclusive knowledge of the cosmos and the earth. The knowledge we gain is inevitably a product of our time. It bears the hallmarks of the always historical interaction between humans and the earth, a process that will never be completed, so it is by definition provisional and unfinished. In that process, humans change, as do the earth and the cosmos. As far as cosmology is concerned, therefore, Nietzsche is a process thinker.

This explains why he referred to his own interaction with the earth as an 'experiment', and sometimes described the history of humankind as a whole as a series of experiments that in the future will merely have to be executed more consciously: 'We are experiments: let us also want to be them!' (D, §453; see also D, §501). For Nietzsche this was a liberating insight. The discovery that life can be allowed to be an experiment encouraged an open attitude in him, a 'yes-saying' to life. 'No, life has not disappointed me. On the contrary, I find it truer, more desirable and mysterious every year – ever since the day when the great liberator came to me: the idea that life could be an experiment of the seeker for knowledge' (GS, IV, §324).[14] Nietzsche's cosmological framework can therefore best be described as interactive. Through the actions of human beings, the earth and the cosmos change, and those changes in turn change humans. In this cosmology, no form of interaction is ruled out, so it even allows for the technological adaptation of the earth. But Nietzsche assesses that interaction like any other, by asking: What does good interaction between humans and the earth involve? That question can be answered only within the framework of a philosophy of the earth. In the next chapter we will see how *Thus Spoke Zarathustra* sketches the early outlines of such a philosophy.

3

Thus Spoke Zarathustra
A tragedy in four acts

Nietzsche's thinking and indeed his life were the product of his personal engagement, which meant a commitment to change. But no one ever seems to have highlighted how profound he believed those changes needed to be and how extensive, nor what Nietzsche encountered in the way of obstacles to that process and how he reacted to them. We know only that by continually drawing upon specific situations and experiences he discovered deeper strata within himself, and ideas about the relationship between humans and the earth that sometimes stimulated and buoyed his enthusiasm for life but mostly frustrated it. Here his questions began, about desire and our experiences of desire: What exactly is desire? Which desires boost our vitality and which restrain it? How does desire operate in animals? What does death mean in nature? Is dying a thing of value? Does it matter if no eternal life follows or is the finality of our end actually a liberation, opening up new ways of looking at how life and death are interlinked? What moral rules are commensurate with our experiences of happiness on earth and where do those rules actually come from? And so on. They are questions addressed by religions and ideologies, of course, but also by philosophers, artists and novelists.

Nietzsche sets to work in his own way on questions of this kind, often drawing upon something concrete, such as an observation, an emotion, an experience, or a reminder of a quotation from Plato or the Bible, or from his own Christian upbringing. The books of his years of wandering – *Human, All Too Human, Daybreak, The Gay Science* – as well as his thousands of notes from the early 1880s are the result. But Nietzsche made no secret of the fact that *Thus Spoke Zarathustra* was the only book in which he truly laid out his vision of the earth and indicated a way of life appropriate to it. All his conjectures and all the efforts he had made up to that point came together there. For him the book represented a breakthrough.

Nietzsche made even less of a secret of the fact that this new insight was closely bound up with that unforgettable experience of 6 August 1881 in the mountains close to Sils Maria, as he was walking from Sils Maria to Silvaplana–Surlej. He alludes to it at the very start of *Thus Spoke Zarathustra*. Having stopped in front of a pyramid-shaped boulder, he had an ecstatic experience that took him out of himself and turned his life upside down. The profoundly disruptive effect lasted for several days. By his own account it revealed to him a radically new understanding of reality. It was a realization that took him by surprise, not the result of reasoning or any other kind of knowledge-based activity. It initially struck him as an existential and emotional earthquake that ran through his whole body. From that moment on, Nietzsche thought and lived differently. It was in the years that followed, between 1882 and 1884, that he wrote *Thus Spoke Zarathustra*, a book in four parts, each of which, his biographers say, he completed in a matter of days, in a trancelike state. It becomes clear just how consequential this episode was when we recall Safranski's observation that Nietzsche's life falls into two parts: the period before and the period after the event at the Surlej boulder. *Thus Spoke Zarathustra* is the book in which Nietzsche wrote most intensely about his personal conception of the relationship between himself and the earth. I now want to turn my attention to it and see whether we can gain a better understanding of the vision it describes.

In my attempt at clarification I have two points of departure. First of all I want to take Nietzsche as seriously as possible in his claim that this one event opened up a totally new perspective in him. What could he mean? The history of science has provided us with the concept of a paradigm shift. A new paradigm suddenly puts all phenomena in a radically different light, while at the same time undermining previously dominant explanations and the prevailing direction of research. Darwin's theory of evolution is one example; it led him to intuit a completely different interpretation of the relationship between humans and nature from that which was generally accepted in his day. We know Darwin arrived at his new insight based on his close observation of variation in nature, after travelling the world for several years on the *Beagle*. He was initially shocked by what he had discovered and for years he resisted making his theory of evolution known to the world, since he expected it to have far-reaching consequences that would not be limited to scientific fields but would extend into matters of morality and religion. Might we be able to explain Nietzsche's experience by using the analogy of a paradigm shift? Might it have been a change in a paradigm whose gravitational centre lies not in scientific knowledge but in our experience of reality, not in a sphere where the process of gaining knowledge is independent and has rules of its own but in the domain where people concern themselves with questions about life and its meaning? Might the key to interpretation be the hypothesis that Nietzsche experienced a radical transformation, which in a flash gave him a completely new view of questions that arise in human beings when they attempt to give direction to their lives? If so, then what Nietzsche experienced as a highly personal event can also be seen as the start of a new kind of knowledge of which the consequences could be sensed intuitively, but which overwhelmed his faculties at that moment.

My second point of departure concerns the style. Stylistically *Thus Spoke Zarathustra* is an extremely unusual book, even in the context of Nietzsche's work. He sometimes described it as a symphony, or mockingly called it 'the fifth gospel'. Zarathustra, the protagonist, speaks in the language of parables, an expressive and metaphorical form. Parables are stories of a kind. Their meaning is not fixed, since they summon a range of associations and emotions. They can therefore be understood or explained in a variety of ways. They require a good listener, in this case someone aware of his or her own lived experience of good and evil, of desire and discomfort, of what is wise and unwise. 'Parables are all names of good and evil,' says Zarathustra early on in the book. 'They do not express, they only hint. A fool who wants to know of them!' (Z, I, 'On the Bestowing Virtue, §1') By speaking only in parables, Zarathustra prompts associations with that other story about the earth, the biblical story of creation, fall and redemption. Jesus too expresses his 'good news' in the form of parables. It is a form that makes nuggets of wisdom about life accessible to our human imaginations. By opting for it, and by then stylizing it further, Nietzsche does not exactly make things easy for us, if only because we are no longer used to thinking in this way. Even at the time it was written, its form represented a break with prevailing ways of thinking. The language of the parable attempts to bring together reality, imagination and life. We need to take time to get used to it if we are to understand it. Another characteristic of the parable is that it addresses the reader directly. All the speech in it and all the scenes that draw us in are conveyed with the intention of raising existential questions in those who are willing to hear, questions about what life on earth means, not in general but for us as individuals. If necessary we will be cast into confusion. All stylistic devices are deployed. It is not a book whose message can be summarized, since the crux is precisely that each reader must undergo it as a personal experience and contemplate what it does to him or her. Readers must each find their own way of reading it, their own personal way of allowing the book to speak to them. This helps to explain the subtitle: *A Book for All and None*.

In my case, *Thus Spoke Zarathustra* spoke to me most clearly when I began reading it as if it were a stage play, a tragedy in four acts. The idea arose after I came upon the following remark a number of times in Nietzsche's notes: 'If Zarathustra wants to move the multitude, he needs to be an actor playing himself' (NF-1881, 12[112]). Nietzsche was a man of the theatre. His best aphorisms are highly theatrical, with an unexpected dénouement, and reading his work therefore provokes much laughter and many tears. It touches us as a good play might. The most successful of the books of his wandering years, *The Gay Science*, makes reference to the art of the troubadours who travelled from place to place in Renaissance times and tried to win people over with music and humour. Troubadours are wanderers too, walkers in the outdoors. But Nietzsche is convinced at a far more fundamental level of the superiority of the theatre as a way of communicating about life. With his knowledge and love of the Greek tragedies – he had read them all and taught on the subject in great detail in his Basel period – he was fully aware of the expressive power peculiar to them and their ability to shine a spotlight on the great questions of human existence, the depths of human desire, the conflicting machinations that arise from them and above all the emotions that tragedies evoke. A tragedy is successful if the audience leaves for home deeply affected and each member of that audience responds in his or her own way. In Nietzsche's view, however, the art of the tragedy met with disaster when Plato and others called into being a new art, namely philosophy, an art that concerns itself with explaining the mystery of life, or with what in life is true and what the mere appearance of truth. To Nietzsche, Plato represented not the start of good philosophy about life but its downfall, because Plato and his central character Socrates – we could perhaps call Socrates Plato's Zarathustra – suggest that it is Plato who can tell the only real, true story about life. If we let this sink in and at the same time realize that Nietzsche was seeking a completely new means of communication, in order to present his intuitions about the earth in such a way that people would be touched by them, it seems obvious that we should regard his principal work *Thus Spoke Zarathustra* as we would a tragedy. It is not an imitation of the classical Greek dramas of course, rather it is a new kind of play, written for modern, urban people, with a message that cannot be deciphered by philosophy, or at least not by what since Plato's day has been called philosophy. Might it not be that with *Thus Spoke Zarathustra* Nietzsche was looking for a new theatrical form, full of prophetic–poetic drama, harbouring all the conflicts he was experiencing himself in his thinking and life, of which the profundities cannot easily be put into words? Is it not possible that Nietzsche wrote a tragedy of his own in order to present the vision of the earth that had come to him at Sils Maria like a bolt of lightning?[1]

These thoughts gave me the idea of presenting *Thus Spoke Zarathustra* in these pages as a stage play. Let me take you with me to an open-air theatre, the 'market place' where people have gathered because, as Nietzsche explains in the prelude to the play, 'it had been promised that a tightrope walker would perform', balancing on a rope strung between animal and overman. It is a play in four acts, each composed of ten scenes. From each act I have selected two scenes and at the start of each act I offer a brief explanation.

Before the curtain rises, a remark about one feature that returns in all four acts. When it came to the Greek tragedies, Nietzsche always had a special fondness for the Greek god Dionysus, the god who puts lust on stage as the deepest motivation of all living things. In *Thus Spoke Zarathustra* the theme of drive and desire, of lust for life, is an obvious leitmotiv throughout the piece. What the earth and humankind have in common, in Nietzsche's view, is the dynamism of primal desire as a driving force in life. He will stage his vision of the earth as an event in which we once again learn to experience 'desire' and 'earth' as aligned, and in which we adapt the earth to make it a *Lustgarten* or 'pleasure garden' that is open to the future. But the route from the city to this new concept of desire will for modern human beings turn out to be full of misunderstandings, traps and hazards. The words 'drive', 'desire' and 'lust' gradually take on new connotations that we need to make our own.

Act One: On the road with Zarathustra

Open your theatre-eye, the great third eye which looks out into the world through the other two! (D, §509)

Scene One: Learning from the animals and looking at the children

The curtain rises. There is sand on the stage, yellow desert sand. Zarathustra enters with a camel, leading it on a rope. The camel is carrying heavy packs. They walk back and forth across the sand in silence. Then Zarathustra makes the camel kneel, whispering, 'Kneel. Camels are good at that!' He loads a few more bags onto the back of the camel and orders it to stand up. Again they walk across the sand. Then he stops, looks at us and announces that he intends to relate a parable, challenging us to divine its meaning.

> Three metamorphoses of the spirit I name for you: how the spirit becomes a camel, and the camel a lion, and finally the lion a child. To the spirit there is much that is heavy; to the strong, carrying spirit imbued with reverence. Its strength demands what is heavy and heaviest. What is heavy? thus asks the carrying spirit. It kneels down like a camel and wants to be well loaded. What is heaviest, you heroes? thus asks the carrying spirit, so that I might take it upon myself and rejoice in my strength.[2] (Z, I, 'The Speeches of Zarathustra: On the Three Metamorphoses')

Zarathustra then leaves the stage with his camel.

From backstage we hear a terrifying roar. Sure enough, Zarathustra now comes onto the stage with a lion, a lion walking free, not on a rope. The lion roars again. Zarathustra goes to sit next to the lion and turns to face it. The lion has a superior look in its eyes, standing there proud and strong. Then Zarathustra says,

> My brothers, why is the lion required by the spirit? Why does the beast of burden, renouncing and reverent, not suffice? To create new values – not even the lion is capable of that: but to create freedom for itself for new creation – that is within the power of the lion. To create freedom for oneself and also a sacred No to duty: for that, my brothers, the lion is required. To take the right to new values – that is the most terrible taking for a carrying and reverent spirit. Indeed, it is preying, and the work of a predatory animal. Once it loved 'thou shalt' as its most sacred, now it must find delusion and despotism even in what is most sacred to it, in order to wrest freedom from

its love by preying. The lion is required for this preying. (Z, I, 'The Speeches of Zarathustra: On the Three Metamorphoses')

While we are still trembling at the strength of the lion and its roar, hoping that the animal will not leap into the audience and grab one of us, a child perhaps, Zarathustra orders the lion to leave. It leaves of its own accord. As we are recovering from the shock we hear new noises from backstage. There is singing, in a child's voice. A child runs onstage, sits down in the sand and starts playing, building a sandcastle. Then it breaks off and starts on something else. A happy child, plainly; a child with a lust for life. A little later Zarathustra appears, sits down next to the child and plays with it. He too is happy, it would seem. He looks at us and says,

> But tell me, my brothers, of what is the child capable that even the lion is not? Why must the preying lion still become a child? The child is innocence and forgetting, a new beginning, a game, a wheel rolling out of itself, a first movement, a sacred yes-saying. Yes, for the game of creation my brothers a sacred yes-saying is required. The spirit wants its will, the one lost to the world now wins its own world. Three metamorphoses of the spirit I named for you: how the spirit became a camel, and the camel a lion, and finally the lion a child.' (Z, I, 'The Speeches of Zarathustra')

Then they dance together, and leave an empty stage behind them.

Moral consciousness takes many forms. That theme is the leitmotiv of the scenes that follow. Zarathustra distances himself from the manifestations he despises: the 'hinterworldlings' who still need a god and a heaven behind the everyday world, 'the despisers of the body', 'the preachers of death', 'the flies of the market place' and all the dominant morality about chastity, friendship and pity. We need to learn to think differently about all these forms of moral consciousness from the way ethics, religion, philosophy and science have taught us to think up to now.

In the final scene of Act One, the perspective on the earth is formulated in positive terms for the first time.

Final Scene: Remain faithful to the earth

On the stage is Zarathustra, who has prepared himself for a journey into the mountains. In the background we can see rows of houses. 'When Zarathustra had taken leave of the city, . . . many who called themselves his disciples followed him, and they provided him escort. Thus they came to a crossroads;

then Zarathustra told them he wanted to walk alone now, for he was a friend of walking alone' (Z, I, 'On the Bestowing Virtue, §1').

> Here Zarathustra was silent for a while and looked with love at his disciples. Then he continued to speak thus – and his voice had transformed. 'Remain faithful to the earth, my brothers, with the power of your virtue! Let your bestowing love and your knowledge serve the meaning of the earth! Thus I beg and beseech you. Do not let it fly away from earthly things. . . . Like me, guide the virtue that has flown away back to the earth – yes, back to the body and life: so that it may give the earth its meaning, a human meaning! In a hundred ways thus far the spirit as well as virtue has flown away and failed. Oh, in our body now all this delusion and failure dwells: there they have become body and will. . . . Let your spirit and your virtue serve the meaning of the earth, my brothers: and the value of all things will be posited newly by you! Therefore you shall be fighters! Therefore you shall be creators! Knowingly the body purifies itself; experimenting with knowledge it elevates itself; all instincts become sacred in the seeker of knowledge. . . . There are a thousand paths that have never yet been walked; a thousand healths and hidden islands of life. Human being and human earth are still unexhausted and undiscovered. . . . You lonely of today, you withdrawing ones, one day you shall be a people: from you who have chosen yourselves a chosen people shall grow – and from them the overman. (Z, I, 'On the Bestowing Virtue, §2')

In *Thus Spoke Zarathustra* the word 'human' usually refers to the figure of the modern person who has swallowed whole the optimistic belief in progress that emanated from the elite of the time. This self-image is denounced by Zarathustra. Modern people must discover themselves as transitional figures anticipating a different culture, Zarathustra tells his listeners. We are 'a bridge and not a purpose'. We must set out in search of a new way of dealing with the earth, a journey that Nietzsche describes as 'the steps to the overman'.

The overman[3] marks a decisive turning point in Zarathustra's message, presenting new prospects for human development. Disciples setting off towards the overman step out of the modern culture of which they are part. Nietzsche directly links the inevitable loneliness they experience with the promise that they will form a new people, a people connected with the earth. By setting out together, over several generations, they will reach the overman. So the overman and the earth belong together. In their mutual involvement, both find their destiny. To mark this insight, Zarathustra later announces the celebration of the 'great noon'. 'And that is the great noon, where human beings stand at the midpoint of their course between animal and overman and celebrate their way to evening as their highest hope: for it is the way to a new morning' (Z, I, 'On the Bestowing Virtue, §3'). Evening and morning, downfall and new beginning, and not once but repeatedly – that is how Zarathustra describes the metamorphosis he requires of human beings. In the third act the same turning point will be described once more, in new words: 'It was there too that I picked up the word "overman" along the way, and that the human is something that must be overcome, that human being is a bridge and not an end; counting itself blessed for its noon and evening as the way to new dawns' (Z, III, 'On Old and New Tablets, §3').

The scene 'Remain Faithful to the Earth' ends the first act and places all the parables that went before it in a new perspective. The purpose and direction of our searching lies with the earth. The way there is laid out according to that basic insight. Yet the new values that we shall discover, experimenting as we go, are only provisional, penultimate values, because the search will go on for a long time, for generations. 'There are a thousand paths that have never yet been walked.'

After Zarathustra has spoken his last lines in Act One, he sends his disciples away. He does not want them to undertake this quest purely because he has told them to. Each of them in their own way must discover a vision of the earth and make it their own. 'You say you believe in Zarathustra? But what matters Zarathustra! You are my believers, but what matter all believers! You had not yet sought yourselves, then you found me. All believers do this; that's why all faith amounts to so little. Now I bid you lose me and find yourselves' (Z, I, 'On the Bestowing Virtue, §3').

Act Two: The game of power

The central theme of the second act is power. What power game is typical of the earth? What powers and powerful figures have people created up to now? Furthermore, does a new vision of the earth change our view of what it means to exercise power? What kinds of power play and leadership contribute to life?

When the curtain rises we see a very different decor: the sea, and in the distance two islands. Above one is written 'Blessed Isle' and above the other

'Fire Hound Isle'. The latter is a gloomy looking place with a volcano belching smoke. The blessed isle bathes in the light of the sun. Why these two islands? Why the sea? Let me explain.

In Sorrento Nietzsche could see two islands from his window, the island of Ischia and next to it the volcano Vesuvius,[4] sticking up above the sea's surface. Ischia became for Nietzsche the symbol of the blessed isle, a familiar theme from classical philosophy that draws upon a tradition more than three thousand years old. The myth of the blessed isle was known to all the peoples around the Mediterranean. Greek poet Hesiod was the first to use the expression 'blessed island' and he writes that Zeus sent the race of heroes there, those who had escaped death on the battlefield. In his teaching in Basel, Nietzsche paid a great deal of attention to these passages, as is clear from the marginal notes he made in his own copy of the text. In the time of Hesiod the island of Ischia was indeed referred to as 'that blessed isle'. It was the westernmost Greek settlement, at the very edge of the known world. Some scholars even claim it was where Odysseus and Nausicaa met, which is quite possible when we consider that not long ago Nestor's cup was found there, with one of the oldest Greek inscriptions ever discovered. From a geographical and climatological point of view too, the island still fits its ancient descriptions. Its soil is highly fertile and scattered with vineyards, olive trees, meadows and fields of grain.

The fire hound symbolizes the smoking Vesuvius that covered the city of Pompeii with its lava, mummifying people and toppling all the statues. Nietzsche made the fire hound the symbol of the state and especially of those political leaders who think that through revolutionary and volcanic violence they can lay out the road to the future. The state, a modern invention, has in Nietzsche's view grown into an exercise of power that ultimately threatens to suffocate the people. He repeatedly calls the state 'the coldest of all cold monsters' (Z, I, 'On the New Idol'). He was suspicious of the ease with which Hegel placed law and reason on an equal footing in his concept of *Staatsräson* or 'national interest'. State power is open to misuse, Nietzsche believes. Several decades later, the state as an institution inspired Kafka to write about the totalitarian exercise of power. We need to keep these historical facts in mind if we are to understand why one of the islands is labelled 'Blessed Isle' and the other 'Fire Hound Isle'.

Scene One: To the islands

Zarathustra comes onstage and hurries across the sea to the blessed isle and there finds once again a number of disciples. After a while he starts to talk about what he calls the overcoming of himself. Again he speaks in parables.

> I pursued the living, I walked the greatest and the smallest paths in order to know its nature. . . . However, wherever I found the living, there too I heard the speech on obedience. All living is an obeying. And this is the second thing that I heard: the one who cannot obey himself is commanded. Such is the nature of the living. This however is the third thing that I heard: that commanding is harder than obeying. And not only that the commander bears the burden of all obeyers, and that this burden easily crushes him. . . . Hear my words you wisest ones! Check seriously to see whether I crept into the very heart of life and into the roots of its heart! Wherever I found the living, there I found the will to power; and even in the will of the serving I found the will to be master. . . . And this secret life itself spoke to me: 'Behold,' it said, 'I am that *which must always overcome itself.* (Z, II, 'On Self-Overcoming', italics in original)

On the blessed isle, Zarathustra sets his disciples on the road towards life, which turns out to feature an all-pervading desire for a will to live, a dynamic that manifests itself as commanding and obeying. On this idea Nietzsche constructed his parable and in doing so he stressed two things: first, obeying and commanding are core activities of life and, secondly, commanding is more difficult than obeying, because a person who commands bears the burden of having to find a direction in the process of evolution. A person who obeys has only to follow.[5] 'Commanding' and 'obeying' are metaphors. Later on in the book, Nietzsche also uses words like 'creating', 'striving' and 'yes-saying', and refers to being at one and the same time 'weigher' and 'scale'. The blessed isle provides the right environment in which to track down the will to live and give it a voice. Those who take up the challenge have to stick close to all expressions of life, listening all the time, and ask themselves what the impulse to command and to create allows itself to be guided by, and why this too is a form of obedience. 'What persuades the living to obey and command, and to still practice obedience while commanding?' (Z, II, 'On Self-overcoming') This is the question with which Zarathustra leaves his disciples.

The central thought in Scene One is: suppose that with this vision of life you wanted to be a leader, Where would you lead people? What is the nature of leadership? Or does the earth itself occupy the role of leader in life and do we people obey?

Nietzsche was fascinated by the dynamic of commanding and obeying that he saw everywhere in nature. The lion, the eagle and all those other predators, and herd animals too, made him think about how vital commanding and obeying were in life in general. In the natural life around him lay the step up to the second question that occupied him: Which expressions of commanding and obeying contribute to the vitality of human beings? It was clear to him that not all forms of human dominion and obedience contribute to it. To find the right ones, Zarathustra advises us to be receptive to the dynamic of commanding and obeying that nature demonstrates, and from there to think afresh about social and political vitality. He returned to these questions in later works with telling titles including *Beyond Good and Evil: Prelude to a Philosophy of the Future* and *On the Genealogy of Morals: A Polemic*.

Scene Two: The fight with the fire hound

Then Zarathustra flies over the sea to the other island, 'on which a fiery mountain smokes continually' (Z, II, 'On Great Events'). There he has a conversation with the fire hound.

'"The earth," he said, "has a skin; and this skin has diseases. One of these diseases for example is called "human being". And another of these diseases is called "fire hound"; about him people have told each other many lies and allowed themselves to be lied to much' (Z, II, 'On Great Events'). The fire hound symbolizes the state and the ways in which modern state power is dominant.

Zarathustra calls the fire hound out of the depths of the earth, curses it and reproaches it, saying,

> At best I could regard you as the ventriloquist of the earth. . . . You've learned more than enough about bringing mud to the boil. . . . Like you yourself the state is a hypocrite hound; like you it likes to speak with smoke and bellowing – to make believe, like you, that it speaks from the belly of things. For it wants absolutely to be the most important animal on earth, this state; and people believe it, too.' (Z, II, 'On Great Events')

In opposition to this political character, Zarathustra presents his own vision of political power:

> And just believe me, friend Infernal Racket! The greatest events – these are not our loudest, but our stillest hours. Not around the inventors of new noise does the world revolve, but around the inventors of new values; *inaudibly* it revolves. And just confess! When your noise and smoke cleared, it was always very little that had happened. What does it matter that a town becomes a mummy and a statue lies in the mud! (Z, II, 'On Great Events', italics in original)

The fire hound responds by behaving 'as though out of his mind with envy'. 'So much steam and so many horrid voices emanated from his throat that [Zarathustra] thought he would choke to death from anger and envy.' Laughing, Zarathustra says,

> You are angry, fire hound, therefore I am right about you! And so that I also remain right, hear now about another fire hound, one who really speaks from the heart of the earth. His breath exhales gold and golden rain – thus his heart wants it. What are ash and smoke and hot slime to him! . . . But gold and laughter – that he takes from the heart of the earth; for you should know – *the heart of the earth is made of gold.* (Z, II, 'On Great Events', italics in original)

What are we to conclude from these two scenes? Is it that human beings cannot lead themselves, let alone a whole people or a state, if they are not capable of understanding deep down what 'life' involves, what 'earth' and 'being in contact with the earth' involve? Do not trust politicians who glorify other values. Real change does not come as the result of political violence and with much 'smoke and bellowing'. Change is the fruit of silent revolutions that bring about a profound alteration in the orientation of values.

After these scenes Zarathustra again leaves his disciples, because his 'mistress' wills it so. She commands him: 'Return to your solitude.' He responds with tears, but obeys. 'When Zarathustra had spoken these words he was overcome by the force of his pain and the nearness of parting from his friends, so that he wept out loud; and no one was able to comfort him. At night, however, he went away alone and left his friends' (Z, II, 'The Stillest Hour').

Act Three: On dreams, time and eternity

On stage we see an urban district in the distance, with newly built houses and a road leading into the mountains. Zarathustra is again about to go into the mountains, but first he takes a last look at the city and the rows of dwellings, and concludes that everything there is becoming smaller and smaller. 'This is because of their teaching on happiness and virtue' (Z, III, 'On Virtue That Makes Small, §2'). When he gets into the mountains he is more alone than ever. The loneliest of all lonely walks is about to begin. 'I am a wanderer and a mountain climber, he said to his heart. I do not like the plains and it seems I cannot sit still for long. And whatever may come to me now as destiny and experience – it will involve wandering and mountain climbing: ultimately one experiences only oneself' (Z, III, 'The Wanderer').

Zarathustra is now about to confront himself at an even deeper level. His mortality and his dreams will have a central role to play. He will pass through extreme experiences and in dreams taste both the positive and the negative sides of life and passion. He also has to struggle with the 'spirit of gravity'. At the end Zarathustra becomes sick, of himself and of humankind. That is the moment when the animals start comforting him by talking to him about the garden and the earth. Again I want to present two of the scenes in this act, with two themes that we have not looked at before. The first concerns Zarathustra's dream, the second is about the experience of time and eternity.

Scene One: My dreams

Zarathustra is standing on the top of a high mountain, alone.

> In a dream, in the last dream of morning I stood today on a foothill – beyond the world, holding a scale, and I *weighed* the world. . . . My dream, a daring sailor, half ship, half whirlwind, silent as butterflies, impatient as falcons: how did it have the patience and while today for world-weighing? . . . How certain my dream looked upon this finite world, not inquisitively, not curiously, not incuriously, not fearfully, not beseechingly: As if a plump apple offered itself to my hand, a ripe golden apple, with cool soft velvety peel – thus the world offered itself to me. As if a tree waved to me, a broad-limbed, strong-willed tree, bent as a support and even as a footrest for the weary traveller: thus stood the world on my foothill. . . . A humanly good

thing the world was for me today, of which so much evil is spoken! How do I thank my morning dream for allowing me to weigh the world early this morning? (Z, III, 'On the Three Evils, §1', italics in original)

The dream brings comfort and a far more positive feeling about the world than Zarathustra generally manages to muster in daytime. He wants to learn from the dream and therefore looks at the world as a dreamer during the day as well.

> In order to do by day what it does, and to imitate it and learn its best, I now want to place the three most evil things on the scale and weigh them humanly well. . . . What are the three best-cursed things in the world? These I want to place on the scale. Sex, lust to rule, selfishness: these three have been cursed best and slandered and lied about most so far – these three I want to weigh humanly well. (Z, III, 'On the Three Evils, §1')

Before Zarathustra starts weighing, he chooses two witnesses to represent the earth. They must observe and check whether his weighing is 'earthly'. The first is the rolling sea, one of Nietzsche's symbols of the earth. Zarathustra holds his scale over the sea where it meets the land. The second witness is a tree: 'You, you hermit tree, you strongly fragrant, broadly vaulted tree that I love!' (Z, III, 'On the Three Evils, §1'). Then he begins to weigh sex, lust to rule and selfishness. 'Now the scale stands balanced and still: three weighty questions I threw into it, three weighty answers are borne by the other pan' (Z, III, 'On the Three Evils, §1').

After each weighing, the outcome is given. We hear which passion is beneficial and which a torment, and above all for whom it is a torment. In the spirit of Dante's *Inferno*, sex, the acts of the despisers of the body, the 'hinterworldly' and all worm-eaten wood are weighed. The result? They are roasted in the 'roiling boil oven'. But for free hearts, innocent and free, sex is 'the garden happiness of earth, all the future's exuberant gratitude for the now' (Z, III, 'On the Three Evils, §2').

Then follows the weighing of the lust to rule. For the corrupt rulers 'before whose gaze human beings crawl and cower and drudge' the lust to rule will become 'the searing scourge of the hardest of the hard-hearted . . . the grim gadfly imposed on the vainest peoples' (Z, III, 'On the Three Evils, §2'). But for 'the pure and the solitary' the lust to rule will ascend 'glowing like a love that luringly paints purple bliss on the earth's skies' (Z, III, 'On the Three Evils, §2').

With selfishness, too, the scale reveals the difference between its good and evil forms. To determine what is good and what is evil from the perspective of the earth, Zarathustra appeals three times for the help of the wisdom of the dream and the two witnesses of the earth. His judgement is that sex, lust to rule and selfishness can be deadly forces in people, indeed they have often been so, but at root they are vital strivings, indispensable in the development that will lead to the overman. Their value and power depends on what is aspired to in them. The answer to the questions 'What for?' and 'To what benefit?', which Nietzsche asks of all human motives, is a determining factor in his value judgement.

Those with some knowledge of philosophy will notice here a reference to the moral philosophy of Immanuel Kant. It was he who called sex, lust to rule and selfishness the three greatest human evils and located them in physicality. Kant could not think of any positive manifestations of them. To him they were wrong by their very nature. Nietzsche has them weighed anew.[6] In doing so, Zarathustra focuses not on Kant but on people who in their self-conceit know 'what is good and evil for humanity' (Z, III, 'On Old and New Tablets, §2'). This delusion disturbs Zarathustra immensely. 'What is good and evil, *no one knows yet* – except for the creator! He, however, is the one who creates a goal for mankind and gives the earth its meaning and future: This one first *creates* the possibility *that* something can be good and evil' (Z, III, 'On Old and New Tablets, §2', italics in original). Then his gaze full of desire wanders 'off into distant futures not yet glimpsed in dreams, into hotter souths than any artist ever dreamed of; there, where dancing gods are ashamed of all clothing' (Z, III, 'On Old and New Tablets, §2'). He goes on: 'It was there too that I picked up the word "overman" along the way, and that the human is something that must be overcome, that human being is a bridge and not an end' (Z, III, 'On Old and New Tablets, §3').

Here, then, Zarathustra allows himself to be led by a source of insight other than reason. Drawn by his desire for a will to live, he experiences his dreams as guides. They open a different perspective on life and above all on what Kant and subsequently Enlightenment morality had determined to be good and evil. A dream connects the day side and the night side of life and unlocks a creative dimension that morality in its highest rational form was unable to open.

Scene Two: Time and eternity

Again a new decor: now seven colossal seals stand on a bare and empty stage. Above them is written, 'The Seven Seals (Or: the Yes and Amen Song)' (Z, III). Those familiar with the Bible will associate the stage set with the seven seals from the Revelation of St. John, Nietzsche's favourite evangelist. In that book the seven seals hold the secrets of the seven calamities that will precede the Last Judgment and announce its coming. The Last Judgment marks the end of time and opens the door to eternal life. So what do the seven seals hide here?

A woman comes onto the stage. She goes over to stand next to the seals and breaks them one by one. Each of the seals reveals an aspect of a new experience of time and eternity. Within Zarathustra's seven seals is the way to an earthly experience of being temporary. The eternal does not come after time but is an experience within time, a time-experience of life, an experience of joy, to be precise. 'Yet all joy wants eternity – wants deep, wants deep eternity,' Nietzsche writes (Z, III 'The Other Dance Song, §3'). When the seal is broken, Zarathustra cries out, 'Never yet have I found the woman from whom I wanted children, unless it were this woman whom I love: for I love you, oh eternity!' (Z, III, 'The Seven Seals, §1' and following) This is repeated loudly at the end of each of the scenes with the seals, as a refrain.

The breaking of the seven seals shows how, with the death of god, heaven and the afterlife have been abolished and how this revelation can discourage people and deprive them of their joy in living. Because how can people suffer and endure sickness and frustration without the prospect of an eternal life to come? All those hoping to answer that question are called upon to free themselves from the morality of heaven and enter into life without any accompanying thoughts to comfort them. Anyone who dares to do this will discover that life features a circular passage of time in which beginning and end continually merge. But there is more. The circular movement typical of earthly creatures is permeated with intensities, with ecstatic moments that take humans out of themselves, as well as intense moments of suffering and frustration. A whole range of intensities is specific to the experience of time. Dare to endure those intensities, dare to say 'Yes and Amen' to them and to discover that they are part of life: that is the message of the seven seals. As always, the animals come to Zarathustra's aid and help him to convey his message, which will be painful for many people. '"Oh Zarathustra," said the animals then. "To those who think as we do, all things themselves approach dancing; they come and reach out their hands and laugh and retreat – and come back. Everything goes, everything comes back; the wheel of being rolls eternally. Everything dies, everything blossoms again, the year of being runs eternally' (Z, III, 'The Convalescent').

Zarathustra arrives at the conclusion that we not only have to tolerate everything that is part of the process we call life but must integrate it into our lives. 'Amor fati' (love of fate) or 'the great health' – expressions that Nietzsche uses to characterize this attitude – do not mean that we should embrace and say yes to everything, whatever it might be. From the perspective of the earth they mean that we must learn to embrace 'life' in all its manifestations. The earthly perspective always sees things from the point of view of 'living' and 'wanting to live' but expands this by including downfall and death. To Nietzsche, 'living' and 'wanting to live' are positive values. Death is not a value as such. If separated from life, death is meaningless. But how could life go on if there were no dying, no continual dying? From that perspective Zarathustra wants to do more than tolerate as unavoidable that which from a human, all too human perspective is hard to endure; he is going to love it wholeheartedly and energetically, as life-giving.

Regarding matters of time and eternity, Zarathustra seems to bring to our attention a different rhythm, the rhythm of daybreak, midday sun and evening twilight, a rhythm of the animals and plants, and of walkers, who live with the sun, moon and stars. It is a rhythm that repeats continually and is experienced intensely in one new form after another by anyone who is open to it.

Again the final scene seems to put all the preceding scenes of the third act into perspective.

Act Four: The steps to the overman

Many 'moons and years' have passed. Zarathustra sits on a rock in front of his cave on the mountain, looking out 'upon the sea and beyond twisting abysses' (Z, IV, 'The Honey Sacrifice'). His hair has turned white. His animals walk around him pensively. There are mountains in the distance, where strange figures wander. One of them has written 'soothsayer' on his cloak. There is a sorcerer dancing, and two kings, who are clearly lost, drive an ass before them. The last pope is there too, blind in one eye it would seem. We can also make out a horribly ugly city dweller, fat and lazy, unspeakably dressed, emitting vile smells that keep everyone at a distance, and last of all an apparently intelligent, slim young man who goes about dressed as a beggar. What they have in common is that they are all visibly searching and unhappy in the depths of their hearts. That is why they have gone into the mountains, into Zarathustra's realm, but they wander about there directionless. Zarathustra has sympathy for them, calls them 'higher men' for their courage and hears their cry for help. He goes up to them and starts a conversation with each of them about what inspires them. We will witness two of these meetings.

Scene One: The meeting with an old man

Zarathustra hurries down. He sees a tall man dressed in black, seated, with a pale and haggard face. Zarathustra is deeply touched.

"'Oh no,' he spoke to his heart, "there sits depression in disguise, and its looks remind me of priests: what do *they* want in my kingdom?"' (Z, IV, 'Retired', italics in original) Then the old man speaks.

> 'Whoever you may be, you wanderer,' he said, 'help a lost, seeking old man who could easily come to harm here! This world here is foreign to me and far off, I even heard wild beasts howling; and the one who could have offered me protection, he himself no longer exists. I sought the last pious human being, a saint and a hermit who alone in his woods had not yet heard what the whole world today knows.' '*What* does the whole world know today?' asked Zarathustra. 'This perhaps, that the old God no longer lives, the one in whom the world once believed?' 'You said it,' answered the old man gloomily. 'And I served this old God until his final hour. But now I am retired, without a master, and yet I am not free, nor merry for a single hour unless in my memories. And so I climbed into these mountains to finally have a festival for myself, as is proper for an old pope and church father: for know this, I am the last pope! – a festival of pious memories and divine worship.' . . . Thus spoke the oldster and he looked with a sharp eye at the man who stood before him; but Zarathustra grasped the old pope's hand and regarded it admiringly for a long time. . . . And with his gaze he penetrated the thoughts and secret thoughts of the old pope. (Z, IV, 'Retired', italics in original)

After the pope has explained how he lost his faith in his god, he goes on talking: "'For our three eyes only,' said the old pope cheerfully (because he was blind in one eye), 'in matters of God I am more enlightened than Zarathustra himself – and am permitted to be'" (Z, IV, 'Retired'). He says angrily that eventually this god 'became old and soft and mellow and pitying, more like a grandfather than a father, but most like a wobbly old grandmother' (Z, IV, 'Retired'). At that Zarathustra interrupts him:

> Did you see *that* with your own eyes? It certainly could have happened that way; that way, *and* another way too. When gods die, they always die many kinds of death. . . . He failed at too much, this potter who never completed his training! But that he avenged himself on his clay formations and his creations because they turned out badly for him – that was a sin against *good taste*. In piousness too there is good taste; *it* said at last: 'Away with *such* a god! Rather no god, rather meet destiny on one's own, rather be a fool, rather be a god oneself!' (Z, IV, 'Retired', italics in original)

Having said these words, Zarathustra invites the old pope, who has visibly brightened, to go up with him to his cave in the mountains. He will invite all the other searchers there too. That is how he closes every meeting. It is striking, incidentally, that Zarathustra approaches all these figures with great compassion. This seems to go against what we know about Nietzsche's attitude to sympathy and pity, about which he is extremely negative. He sees pity as behaviour that makes humans small and keeps them small. It arises from a mistaken understanding of love for fellow humans and prevents both the giver and the receiver from truly following their life force. That is one side of Nietzsche's teaching on pity. But from the perspective of the earth, and on the way to the overman, it turns out that compassion can be a highly valuable attitude to life. Compassion is mixed here with irony and warm ridicule.

With renewed courage, Zarathustra then approaches the others. After meeting with 'the ugliest human being', 'the unspeakable one', the figure who embodies the lowest of the sentiments that modern life arouses in people – disgust, avarice, superficiality, gluttony, aggression – he sets out to meet a young beggar.

Scene Two: The meeting with the young man, a voluntary beggar

> When Zarathustra had left the ugliest human being, he was freezing and he felt lonely; after all, so much that was cold and lonely went through his

mind, to the point where even his limbs grew colder because of it. . . . Then all at once his mood became warmer and more cordial. 'What happened to me?' he asked himself, 'something warm and lively refreshes me, something that must be close to me. Already I am less alone; unknown companions and brothers roam around me, their warm breath touches on my soul.' But when he peered about himself and searched for the comforters of his solitude, oddly enough, it was cows huddled together on a knoll; their nearness and smell had warmed his heart. . . . When Zarathustra was quite near them he heard clearly how a human voice spoke from the midst of the cows; and evidently they had all turned their heads toward the speaker. . . . 'What are you seeking here?' cried Zarathustra, astonished. 'What am I seeking here?' he answered: 'The same thing you seek, you trouble maker! Namely happiness on earth. But for that I want to learn from these cows. And you should know, I've already persuaded them half the morning, and just now they wanted to tell me for sure. Why do you have to disturb them? Unless we are converted and become as cows, we will by no means enter the kingdom of heaven. For there is one thing that we ought to learn from them: chewing the cud. And truly, what profit is it to a man if he gains the whole world, and did not learn this one thing, chewing the cud: what would it help? He would not be rid of his misery – his great misery: which today is called *nausea*.' (Z, IV, 'The Voluntary Beggar', italics in original)

The voluntary beggar represents humans who have relinquished the riches modern life has to offer and are ashamed of the greed that accompanies wanting to become rich. He has been driven to the poorest by feelings of nausea inspired in him by the rich, by 'the convicts of wealth who cull their advantage out of every dustpan, with cold eyes, horny thoughts; for this mob that stinks to high heaven, for this gilded, fake rabble, whose fathers were pick-pockets or vultures or rag pickers' (Z, IV, 'The Voluntary Beggar'). Between the lines, this character evokes associations with the Christ figure, who laid aside all divine power and went about as a beggar, in search of happiness on earth.

> Thus spoke the mountain preacher and then he turned his own gaze on Zarathustra – for till now his gaze hung lovingly on the cows – then, however, it transformed. 'Who is this with whom I speak?' he cried, startled, and jumped from the ground. 'This is the man without nausea, this is Zarathustra himself, the one who overcame great nausea, this is the eye, this is the mouth, this is the heart of Zarathustra himself.' And while he spoke thus he kissed the hands of the one to whom he spoke, and tears streamed from his eyes, and he behaved quite like someone to whom a precious gift and treasure falls unexpectedly from heaven. The cows, meanwhile, watched

all of this and were amazed. . . . 'Well then!' said Zarathustra. 'You should also see *my* animals, my eagle and my snake – their equal exists nowhere today on earth. Look, there the path leads to my cave; be its guest tonight. And speak with my animals about the happiness of animals.' (Z, IV, 'The Voluntary Beggar', italics in original)

When all the meetings are over and all the 'higher men' have walked up the mountain to Zarathustra's cave, he goes there himself and invites them for a 'last supper'. During the meal he talks with them further and they discuss what motivates them as higher men. But meanwhile Zarathustra is forced to conclude that they have not understood what he means by the overman. At the end of the evening, when his guests' eyes begin to shut, he slips outside.

'Oh clean fragrance around me,' he cried out, 'oh blissful stillness around me! But where are my animals? Come here, come here my eagle and my snake! Tell me, my animals: these higher men all together – do they perhaps not *smell* good. Oh clean fragrances around me! Only now do I know and feel how I love you, my animals.'[7] (Z, IV 'The Song of Melancholy, §1', italics in original)

When Zarathustra reaches the end of his search and has seen for himself that neither the people, nor his disciples, nor the 'higher men' he met on his journey have understood much about the dynamics of life, of the earth and the overman, he remains there alone. Once again it is the animals that receive him, ecstatic with joy. 'The eagle and snake pressed up against him as he spoke these words, and they looked up at him. In such a manner the three of them together sniffled and sipped the good air. For the air here outside was better than among the higher men' (Z, IV, 'The Song of Melancholy, §1'). Later, 'the doves flew back and forth and lighted on his shoulders and caressed his white hair and did not tire of tenderness and jubilation' (Z, IV, 'The Sign'). Even the lion, a symbol of strength and will in nature, 'pressed its head against his knee and did not want to leave him out of love, acting like a dog that finds its old master again, . . . licking the tears that fell onto Zarathustra's hands, roaring and growling bashfully. Thus acted these animals. All this lasted a long time, or a short time: for, properly speaking, there is *no* time on earth for such things' (Z, IV, 'The Sign', italics in original).

Then the curtain falls for the final time

Recap: The great transformation

In *Thus Spoke Zarathustra*, Nietzsche left us more than just a book. It is an experience, an overwhelming experience in fact, leaving us somewhat dazed and bemused, with many different emotions, impressions and questions. The path that Zarathustra climbs offers us a glimpse of all the things that will change if modern humans truly undertake a turn towards the earth in their lives and their thinking. He undertakes that journey for us. Every next step, every new scene is full of unexpected associations and thought-provoking references, with questions, commands and exclamations. We are led into a totally new way of looking at the life around us and at ourselves. We have been shown that all life involves a game of commanding and obeying, of creating and taking orders. Through Zarathustra's dream we come into contact with the deepest layers of life, which prompt broader and milder judgements. The suffering that humans cannot avoid is illuminated by means of a vision of transience and eternity that the animals summon up. Finally, all humans who are sincerely seeking are invited to come to Zarathustra's cave, high on the mountain, where the spectacle of life on earth can be seen in its full extent.

That ultimately not one of the people Zarathustra meets understands him has been explained by Paul van Tongeren as the ultimate proof that Zarathustra's project has failed.[8] I prefer to draw an alternative conclusion, namely that it was impossible for modern humans to understand Zarathustra's project because Zarathustra unlocks a radically new orientation of humans to the earth. Hence the subtitle *A Book for All and None*. The overman appears only on the distant horizon, and therefore Zarathustra could feel at home only with the animals. This too makes *Thus Spoke Zarathustra* a tragedy, but one that ends in a euphoric apotheosis that fills all living creatures on earth with intense happiness, all except humans. This outcome suggests what Nietzsche may have meant when he chose as a motto for his philosophy the *Umwertung aller Werte*, or revaluation of all values. For that reason, *Thus Spoke Zarathustra* has been called his most ambitious and profound thought experiment.

At the start of this chapter I suggested that *Thus Spoke Zarathustra* can be thought of as a paradigm shift, a radical transformation in Nietzsche's thinking about the meaning of life. At its core is the way that he makes the living earth the heart of a new philosophy of existence and the art of living well. He allows us to discover the earth as truly immense, as the great stillness, as a reality that cannot be governed, as an overwhelming force, as the source of values and emotions that guide life, as a mystery of a completely new kind. In the philosophical universe, empty after the death of god and filled with nihilistic moods and ideas, Nietzsche presents the earth as the new horizon of life. 'Remain faithful to the earth, my brothers, with the power of your virtue! Let your bestowing love and your knowledge serve the meaning of the earth! Thus I beg and beseech you. . . . Like me, guide the virtue that has flown away back to the earth – yes, back to the body and life: so that it may give the earth its meaning, a human meaning!' (Z, I, 'On the Bestowing Virtue, §2'). Can we take the wisdom of the earth, the 'terrasophy' presented here in the form of a stage play, as a guide for a new philosophy? Do the ideas that Nietzsche explores offer points of departure for thinking through the questions raised in our own time about the relationship between humankind and the earth? I will consider these questions in Part II

Intermezzo: Gary Shapiro, *Nietzsche's Earth: Great Events, Great Politics* (2016)

Three reflections

Of late there has been increasing interest in Nietzsche as a 'philosopher of the earth'. Six months after this book was first published in Dutch, Gary Shapiro's *Nietzsche's Earth: Great Events, Great Politics* appeared. It has since become a standard work, both because Shapiro interprets *Thus Spoke Zarathustra* in the light of Nietzsche's other writings and because he explains Nietzsche's philosophy of the earth as political as well as philosophical in conception, a proposal for a radical transformation of the way we inhabit the earth.

In this intermezzo I want to look briefly at a few important insights from Shapiro's book. By examining a number of different themes, I will suggest how my own work relates to Shapiro's political interpretation of Nietzsche's philosophy of the earth.

For Shapiro, 'remain faithful to the earth' is the key concept in Nietzsche's thought, but he interprets the affirmative slant Nietzsche gave it in *Thus Spoke Zarathustra* in the light of the critical and political attitudes to be found in his later works, especially *Beyond Good and Evil*. Shapiro writes that the two books belong together and presume each other like yin and yang:

> Superficially, they are in sharp contrast. Myth, drama, song, and landscape seem very different from analysis, argument, and a detailed focus on the follies of contemporary Europe. At a deeper level, they are indeed the yin and yang of a philosophical vision of human life on the earth and its prospects for the future. (2016, p. 75)

The twinned nature of these books puts Shapiro on a path towards developing 'an alternative to an overly interiorized, subjectivist and "existentialist" reading of Nietzsche', an interpretation of his own in which the earth is defined as a political concept and Nietzsche's work as above all of a political–philosophical nature. He describes how Nietzsche's philosophy of the earth arose out of his sequence of different reactions to Hegel's philosophy. Shapiro systematically zooms in on the contrasts between Hegel's concept of the world and world history and Nietzsche's ideas about the earth. In doing so he succeeds in focusing our attention on the new orientations that Nietzsche brings to the fore. In that light I want to touch upon two matters arising from Shapiro's work that received insufficient attention in the original Dutch edition of this book, and to give my response to them.

The first concerns Shapiro's claim that Nietzsche introduces a new narrative with his vocabulary of the earth. Wherever the word 'earth' appears ('remain faithful to the earth'; 'the meaning of the earth'; 'sacrifice themselves for the earth'; 'an earthly head that creates meaning for the earth'; 'the kingdom of the earth'; 'the garden happiness of earth' and so on) his work can be read as the impetus for a radical new vision. Nietzsche often, almost in passing, weaves his unique vocabulary of the earth into his argument. By bringing these passages to the fore and making their purport visible, Shapiro succeeds in substantiating his claim that they are typical of a new philosophy that Nietzsche elaborates upon in a different way each time.

The fruitful nature of this contrastive means of interpretation becomes very clear around the subject of the nation state, a central concept in Hegel's philosophy. Nietzsche is strongly opposed to the implicit ideology of national unity that Hegel advocates. 'I, the State, am the people' is one of the succinct formulations Nietzsche uses to stigmatize any such identification. Shapiro argues that Hegel's concept of the state is unmasked by Nietzsche as inevitably devaluing the unique individuality of human beings in favour of efforts to establish a uniform and homogenous identity. Statist thinking justifies the establishment of national borders, coupled with territorial claims, with sharp dividing lines between inside and out. At an early stage, Nietzsche, with his sensitivity to the earth, tried to undermine this network of statist ideas. He unmasks governments and rulers as representatives of very different interests from those of the earth. He does not regard the leaders of the state – or of the church, which according to Nietzsche is in cahoots with the state – as capable of shaping the relationship between humankind and the earth. We have already seen how strongly Zarathustra expresses this when he calls the leaders of the state 'fire hounds' (Z, II, 'On Great Events').

Shapiro then identifies the alternative lines of thought that Nietzsche explores. As against the imposed identity of the citizen of a nation state, for example, Nietzsche advocates the shaping of people into good Europeans. Encouraged by the cultural mixing that was occurring even in his day, he was in favour of experimenting with new cultural and political hybrid identities. He strongly rejected the idea of reducing peoples to masses that have to be moulded into a homogenous form. According to Shapiro, he argued that we should see peoples as 'multitudes'. They are full of productive possibilities precisely because they constitute experiments in inhabiting the earth. They are mobile, like nomads in their wanderings and migration, and they require this mobility, as well as climatic and environmental changes, if they are to perpetuate the human species (2016, pp. 91–93). It is these multitudes that give birth to hybrid humans, forerunners of the good Europeans of the future, who try out different cultural combinations and syntheses (2016, p. 97). Here Shapiro shows that in his reflections on a new relationship with the state, Nietzsche is inspired not just by the figure of the 'good European' but by nomads and migrants. Shapiro's analyses are extremely productive. They help to throw new light on passages by Nietzsche about the earth that are generally brief and often implicit or hard to interpret.

Shapiro's reconstruction of Nietzsche's philosophy of the earth is not the only possible interpretation, as he himself admits. In this book I want to show that Nietzsche's focus on the earth leads him to develop his thinking in another direction as well. It comes into view when we look at how he lets go of the abstract, rationalizing and universalist concepts of space and time. In opposition to them he places the plurality and richness of local cultural expression. As we saw in Chapter 2, he became fascinated by the influence of climatic differences. How do local communities nestle into the landscape around them in the heat of the tropics? How does that differ from the way people in colder northern regions settle the landscape? Differences in climatic conditions obviously cause people to have different ways of feeding and clothing themselves, but Nietzsche claims they also produce differences in character, in sensitivity and in morality. He therefore rejects Kant's claim that the categorical imperative is universal. In fact he rejects any notion that morality could ever be universal. Nietzsche's conviction that cultures are multifarious and local by their very nature is an important source of inspiration for his concrete, tangible philosophy of the earth. In Chapter 2 I gave some examples. In Chapter 5 I will show how productive this way of thinking can be in our own day.

A second contrast between Hegel and Nietzsche that Shapiro points out concerns their vision of history and of the future. Hegel's philosophy of the state is the coping stone of his ideas about history. The development of humankind is framed within that philosophy as a logical progression from the spirit or logos, the outcome of which is the modern state. Because this rational society, a construct of rights and laws, is the highest form that humankind can reach in Hegel's view, it also constitutes the framework within which the future is interpreted and regulated. In that sense, therefore, the state represents 'the end of history'. Hegel captures historical becoming in a story with a beginning, a middle and an end. Irrespective of what the future may hold, he interprets everything from this perspective.

Nietzsche rejects increasingly radically this totalitarian and teleological vision of world history and the anticipation of the future that is based upon it. He does so in part by giving Hegel's term 'great events' an opposing connotation. Hegel uses it to describe events or acts that play an essential part in the development of the human species, as distinct from the chance, fortuitous events of everyday. Philosophers are to concern themselves with the former, he writes; the latter are the stuff of news, fodder for journalists. Only a philosophical view of history (meaning his own) can distinguish between the two, Hegel claims, and make it possible to discern why some things that happen turn out to be necessary in the course of the development of the spirit. Whereas for Hegel 'great events' bring arrival at the state closer, for Nietzsche 'great events' are those that, rather than filling in the future in advance, break it open. They unlock a radical new future (see Chapter 2). The contrast between these two definitions is clear in a famous passage in *Thus Spoke Zarathustra* in which Zarathustra calls the state a 'hypocrite hound' that merely makes noise and cannot bring about real change. 'And just believe me, friend Infernal Racket! The greatest events – these are not our loudest, but our stillest hours. Not around the inventors of new noise does the world revolve, but around the inventors of new values' (Z, II, 'On Great Events').

The 'infernal racket' refers to the often violent acts with which states force through changes that, Nietzsche believes, are not real changes at all. Radical change happens when a reversal of values takes place. The new horizon of values is referred to here as 'the earth'. Nietzsche writes that 'the heart of the earth is made of gold'. To clarify this radically open vision of the future, Shapiro introduces the term 'futurity'. The question he aims to answer is as follows:

> How can we begin not so much to envision specific futures as to incorporate the always indeterminate futurity of the earth into our concept of the political? . . . Those attuned to futurity are open to seizing the gift of fortune, the moments of opportunity, the fleeting moment of great possibilities that the ancients call kairos. (2016, pp. 2–3)

Nietzsche conceives of the present as a possibly decisive moment for a new future that he connects with a new attitude towards the earth. Shapiro uses the term '*kairos*' for this unity in the experience of time. It is distinct from the *chronos* vision of time and temporality, in which all moments are inevitably part of a linear narrative. The logic of *chronos* subjects both personal and political time to the rational regime of world history and is therefore tempting to anyone eager for power. To cultivate an attitude born out of *kairos* we need to escape from the temptations of *chronos*.

In *Beyond Good and Evil*, Nietzsche goes on to make this attitude a task for 'higher men', for those who want to live a noble life. To seize 'moments of opportunity', to use Shapiro's term, they must think beyond all 'father- and motherlands' and open themselves up to a 'children's land, the undiscovered land in the furthest sea' (Z, II, 'On the Land of Education').

Shapiro believes that the dimension of futurity – epistemological, existential and political – forms the core of Nietzsche's new, future-orientated experience of time and historicity. The *Übermensch* or 'overman', Nietzsche's primary metaphor for the future, is connected to it as well:

> We can say that the earth will be a home for the enigmatic figure of the Ubermensch, a figure defined by its own radical futurity, a futurity set free in the innocence of a becoming without debt (Unschuld des Werdens). Is there something more to be known about the form of the earth's futurity? How could the earth be transformed so as to become the ground or theatre of great events? (2016, p. 135)

Entirely in line with Shapiro's view, in this book I interpret the *Übermensch* as the metaphorical figure of the open future. This 'overman' does not fit into the framework of *chronos*, in which, based on the past and the present, the outlines of the future can be drawn in anticipation. The 'overman' does not represent an 'end of history' in the Hegelian sense. In my interpretation too, we cannot imagine the 'overman', nor does the concept refer to an ideal to which we can already give content. When Nietzsche calls on his disciples to take 'the steps to the overman', he consistently declines to draw the precise outlines of that figure, instead searching for powerful expressions that will get through to readers and set them moving. He uses the language of emotion, of desire, language intended to convey urgency, to awaken hope, remove doubt or overcome putative passivity or mistrust. It is in such language that the attitude of humankind to futurity takes shape. Although the words may sound somewhat melodramatic to modern ears, this linguistic mode was chosen by Nietzsche as the most likely to bring about action:

> Remain faithful to the earth, my brothers, with the power of your virtue! Let your bestowing love and your knowledge serve the meaning of the earth! Thus I beg and beseech you. Do not let it fly away from earthly things.... Like me, guide the virtue that has flown away back to the earth – yes, back to the body and life: so that it may give the earth its meaning, a human meaning! ... Let your spirit and your virtue serve the meaning of the earth, my brothers: and the value of all things will be posited newly by you! Therefore you shall be fighters! Therefore you shall be creators! ... There are a thousand paths that have never yet been walked; a thousand healths and hidden islands of life. Human being and human earth are still unexhausted and undiscovered. ... You lonely of today, you withdrawing ones, one day you shall be a people: from you who have chosen yourselves a chosen people shall grow – and from them the overman. (Z, I, 'On the Bestowing Virtue, §2')

Shapiro is right in thinking that futurity requires that things must be set in motion. By bringing to the fore the language characteristic of summons and exclamation, of hope and desire, and by showing that the animals have an important part to play, I hope to illustrate this dimension. In contrast to Shapiro's reconstruction, the animals are important in my book. Emotions add to alertness (characteristic of futurity) a profound desire for action. Alertness is again given expression, although in a different way, in the following passage from *On the Genealogy of Morals*. It is an entreaty, inspired by fear and hope, that aims to shout down the doubt that can be heard in the background, but it expresses above all Nietzsche's longing for people who want to take steps to the overman.

> But some time, in a stronger age than this mouldy, self-doubting present day, he will have to come to us, the *redeeming* man of great love and contempt, the creative spirit who is pushed out of any position 'outside' or 'beyond' by

his surging strength again and again, whose solitude will be misunderstood by the people as though it were flight *from* reality: where it is just his way of being absorbed, buried and immersed in reality. . . . This man of the future will redeem us, not just from the ideal held up till now, but also from those things *which had to arise from it*, from the great nausea, the will to nothingness, from nihilism, that stroke of midday and of great decision that makes the will free again, which gives earth its purpose and man his hope again, this Antichrist and anti-nihilist, this conqueror of God and of nothingness – he must come one day. . . . But what am I saying? Enough! Enough! At this point just one thing is proper, silence: otherwise I shall be misappropriating something that belongs to another, younger man, one 'with more future', one stronger than me – something to which *Zarathustra* alone is entitled, *Zarathustra the Godless*. (GM §24–25, italics in original)

Finally, and this is my third reflection, Shapiro's book about Nietzsche's philosophy of the earth differs from mine in its aim and intention. Shapiro analyses all of Nietzsche's work and combines a chronological approach with thematic elaborations. His book can be read, he writes, 'as a series of philological commentaries' (2016, p. 18). He presumes on the part of the reader a considerable amount of knowledge of Nietzsche's thought and of contemporary philosophy. Although Shapiro also aims to 'sharpen thinking about contemporary planetary crises and Nietzsche's view of the political by explicating his injunction to give the utmost priority to the question of the direction (*Sinn*) of the earth' (2016, p. 20), this is generally limited to philosophical references.

My book is written with another purpose altogether, namely to provide an introduction to Nietzsche's philosophy of the earth for readers who are unfamiliar or less than fully familiar with his work. I have devoted Part I to achieving that aim. In Part II I will explain how Nietzsche has inspired me in elaborating upon three themes that I regard as important in the context of the current climate crisis: ways of working to develop a personal ecological lifestyle; a revaluation of the connection between local communities and their environments; and the need for a new cosmology, a new wisdom regarding the earth, for which I introduce the term 'terrasophy'.

Part II

'Terrasophy': Guidelines for a philosophy of the future

We live in a very different time from Nietzsche's. We have vastly more knowledge about the earth than was available to the scholars of his day, although we did not truly became aware of the earth's uniqueness until astronauts showed us what our planet looks like from space: a small blue globe surrounded by a wafer-thin biosphere. That image of the earth has been firmly fixed in our minds ever since. More recently, we have been subjected to a steady stream of information about the consequences of human activity on earth and how, over the years, the position of humankind on the planet has changed dramatically. Reports on the subject are so plentiful that we can barely take it all in, and many, myself included, experience the present day as a time of crisis. We feel a need to focus on the situation in which we find ourselves and on what can be done about it.

It was because of these concerns that I reread Nietzsche, in the hope of finding points of departure for dealing with the issues that trouble us today and inspiration in my quest to discover the correct attitude to life. What came home to me above all was that Nietzsche made restoring humankind's bond with the earth a personal experiment, exploring step by step how he could achieve that aim in his own life. I found it striking that this released so much positive energy in him and put him on track to a life that was dominated by questions about the health of mind and body, about vitality and joie de vivre. He became sensitive to the happiness of animals and to the beauty of nature, and his ideas about what could be called a good or a happy life changed radically as a result. Wanting to be in contact with the earth turned out to be another description of wanting to live an authentic life. The positive experiences he gained and his continual battle with everything that frustrated and blocked his progress led to a new way of dealing with philosophical problems, a way that was complex, full of allusions and metaphors that are sometimes more suggestive than explanatory, but that did not avoid fundamental questions. Moreover, using Zarathustra as a mouthpiece, he confronts his readers with questions about the extent to which we are still imprisoned in the modern culture that has shaped us, and calls upon us to take the path towards the earth without deviating and without allowing ourselves to be discouraged. In Nietzschean terms, getting into contact with the earth ultimately means focusing on the future and on future generations. Amid the fears and doom-laden scenarios that are so prominent a part of today's growing ecological awareness, as perhaps they must be, Nietzsche presents new prospects for hope and creativity. The planetary orientation he fosters is both pragmatic and aesthetic, both hedonistic and spiritual. It does not amount to a blueprint for what we need to do, rather it evokes the image of a beckoning perspective that opens up new horizons as people venture further along this road to the future.

Just how radical is the change of direction proposed by Nietzsche becomes clear when we set his approach against the background of the history of philosophy. The various schools of ancient philosophy define their discipline with reference to three interconnected themes: the relationship of human beings to themselves, to other people and to the cosmos. Care for oneself and care for others are placed in the broader framework of the relationship between humankind and the universe. Again and again the question arises of how personal life relates to *physis*, the all-encompassing reality that includes the cosmos. In answering that question, philosophers have always linked two dimensions of knowledge. In ancient philosophy, cognitive knowledge of the self was brought together with pointers on how we should live; knowledge about living together, in the polis or outside it, was linked with instructions on justice and social interaction; and knowledge of nature and the cosmos of which nature is part was connected with reflections on the place that falls to humans within the universe as a whole. This triangle of ethics, politics and physics threw up questions about our identity as humans and produced instructions for a way of life appropriate to our humanity. The cosmic vision, or a view of the world in its entirety, has always been the most difficult to achieve, but schools of philosophy, especially those of Epicurus and the Stoics, provided gymnastic, philosophical and spiritual exercises to train us to have such a view of our own lives and of life with others.

From the Renaissance onwards modern philosophy broke with this basic triangular pattern. It no longer took the cosmos as a coordinating guide for existence but put in its place the individual human being, separated from the larger whole. Not the organization of the universe and the effect it had in nature but individual freedom and the pursuit of independence became the founding principles. People were encouraged to define themselves as distinct from pre-existing structures and to liberate themselves from social frameworks, which were seen more as hindrances than as supports for human self-development and happiness. The umbilical cord between cosmic and human reality was cut, partly on the authority of the scientific discoveries made by Galileo, Copernicus and Newton, who brought about a definitive break with previous cosmologies. The universe was henceforth seen as a neutral reality, indifferent to humans and offering them nothing on which to base their ways of life. Pascal vividly expressed the experience of existential emptiness created by this world view when he wrote *'Le silence eternel de ces espaces infinis m'effraie.'* (The eternal silence of these infinite spaces alarms me.) This vision of humanity as a self-founding starting point of thinking and acting was further developed by the Enlightenment thinkers of the eighteenth century. Throughout the two centuries that followed, it shaped the Western view of the world and the flourishing humanities and social sciences. The outcome was a discourse about human emancipation, human development and interpersonal social justice that has had a huge influence up to our own day. The norms for well-being and dignity that flowed from it proved, however, to have a large ecological footprint. Today we are discovering the price paid by the earth for what modernity as a norm for human development has created. With the arrival of the Anthropocene, this insight has become mainstream.

In relation to these two basic templates, the ancient and the modern, Nietzsche's philosophy takes a position all its own. Nietzsche investigates modernity and the core values on which it rests to assess their hostility to nature. He has his alter ego Zarathustra say, 'The earth has a skin; and this skin has diseases. One of these diseases for example is called: "Human being"' (Z, II, 'On Great Events'). He puts 'human being' between inverted commas, since he is referring to the modern image of humanity as sketched out above. Nietzsche is close to us, because he carries out his philosophical project in discussion with that basic template, which he calls deeply alienating. He is determined to work on a philosophy of the universe, and in particular to make the earth once again a guide to humanity. But he does not return to the cosmology of the ancients, because he does not share their notion that the universe is orderly and rational. Instead his quest is to develop a new philosophy of the earth, beyond both that of the ancient world and that of modern cosmology. The new trail he sets out can best be visualized with the help of the classic triangle found in ancient philosophy: care for oneself, care for others, and the relationship between human beings and the cosmos.

In looking at how humans relate to themselves, Nietzsche does not take as his starting point the modern individual, an abstract figure, stripped of a body. Instead he begins from a concept of the human as an actual, bodily person, product of a long evolution, who through nourishment and the cultivation of the environment is interwoven by many threads with a specific place of residence on earth. The importance of the physical body and of local connections was neglected in the modern paradigm. Fresh consideration of these things therefore becomes an important line of thinking in his new cosmology. Another difference between his philosophy and that of the ancient world and of modernity is that Nietzsche positions the individual as a protagonist, always unique and solitary. He has an affinity with the existential dimension present in ancient philosophy, but he introduces a new subject. In setting out this existential line he is always in conversation with the Epicurean and Stoic tradition of care for oneself from a cosmic perspective.

When it comes to the relationship between humans and the cosmos, Nietzsche introduces an even more radical change of perspective. He does not in the first instance focus his gaze on the universe, the awe-inspiring firmament of sun, moon and stars above the earth. He tilts that perspective by 180 degrees and looks at the earth itself and at life on earth. Instead of a cosmic gaze, looking from the earth at the heavens, in Nietzsche we come upon the gaze of the eagle, forming a view of the earth from high above. His cosmic perspective is on this point close to that of the astronauts who, from beyond the earth, from the cosmos, looked back at the living planet and discovered how all life on earth is interconnected in one vast ecological network. Philosophical thought is guided here not by the unchanging order of the universe but by a desire to fathom the secrets and forces that life on earth contains within it, a desire that helped to shape his view of the planet and inspired the questions he asked in his new research. This radical change of perspective has led me to introduce the term 'terrasophy' for Nietzsche's cosmological undertaking.

Before I elaborate on that term any further, I first want to look at the third cornerstone of the philosophical triangle, that is, the relationship between the individual and other people, which coalesces in the concept of politics. This subject has received a great deal of attention in modern philosophy. The principles of freedom and equality have led us to view social behaviour as a basis for the rule of law and a democratic society, and for the shaping of a universal morality. It is a morality that takes the dignity of all humans as its starting point and makes the right to a life worthy of a human being the basis of a Universal Declaration of Human Rights. This universality is intended to include all members of our species. For the way in which it has done so, however, the price is high, Nietzsche believes, since it makes 'humanity' an abstract quantity, lacking any material, earthly basis. The fundamental principles of freedom, equality and dignity have broken loose from the reality of everyday life. Nietzsche therefore prefers to start from the actual person and the local community. Against the universal perspective – 'all human beings' – that modernity takes as a guiding principle for regulating life on earth he places the multiplicity and complexity of local cultures and local communities. Not universality but the values of plurality and diversity initially come to the fore. They become the foundational values for a new political order and a planetary morality. The implication of Nietzsche's thinking, therefore, is that we need a new framework for the political dimension of society, one that enables us to deepen the political–ethical perspective of modern universal morality and transform it by connecting it to the earth. The concept of terrasophy enables us to explore this threefold change of perspective further.

I define terrasophy as a new direction in philosophical cosmology that makes the relationship between humans and the earth the central concern of our twenty-first-century understanding of ourselves. The current phase of development of life on earth is forcing us to address this issue and think it through. Terrasophy presumes locally situated, physical people who are compelled, and who also desire, to live out their lives together. It contemplates their vital urges, both personal and social, from the perspective of the earth as a whole. Terrasophy connects the view from below, the outlook of the corporeal, personal and local, with the view from above. It is the view from above that gives every person a notion of their place within the larger whole of life on earth. It is from above that the earth appears as a multicoloured reality full of diversity and variety. Terrasophy aims to keep moving back and forth between the personal and local and the planetary perspective, between on the one hand the experience of being intensely and bodily part of life and absorbed by it, and on the other of transcending it from a perspective originating from above. Repeatedly supplementing personal experience with questions and thoughts about the actual life on earth of people elsewhere creates insight into what it means to live on one planet along with other life forms. Despite the prevailing attitudes of his day, Nietzsche realized that we share the earth not just with other people but with all living creatures.

Nietzsche symbolically connected these two perspectives, the personal and the planetary, in the animals that Zarathustra always has with him during his great metamorphosis and regards as his personal advisers: the eagle and the snake. It is impossible to come any more closely and bodily into contact with the earth than a snake does, slithering over the ground. So who can advise Zarathustra better than the snake? And in the whole of the animal kingdom there is no all-encompassing view to rival that of the eagle, scouring the wide, living landscape from on high with its powerful eyesight. For his own meditations, Zarathustra often climbs to the highest rocks, from where he has a broad vista. The challenge posed by terrasophy is to broaden the existential dimension and at the same time to return to the earth the moral perspective of all people, the great achievement of modern morality from which we derive our core values.

In developing a terrasophical framework, therefore, I am guided by the question of how the self, in its existential and physical complexity, can bring together being locally situated with having a perspective on the earth as a whole. I want to explore what the interweaving of the two looks like in our time and what tasks it places before us. Terrasophy also prompts a new way of thinking, one that brings cognitive knowledge back together with morality and with spiritual knowledge of a practical kind, so that they nourish each other and create new openings by their reciprocity. In his quest for a new way of philosophizing, Nietzsche was inspired both by his own experiences and observations and by the life sciences; he also drew upon ancient and more recent philosophical traditions, and on art and literature. The '-sophy' in terrasophy points to a new type of earthly wisdom in which knowledge and intelligence, life skills and scientific insights are interconnected.

Part II of this book is therefore designed to be an exploration, an attempt to supply building blocks for the development of a terrasophy based on the three themes specified above: the relationship of human beings to themselves, to other people and to the cosmos.

In Chapter 4 I explore how Nietzsche's focus on the personal point of view and his existential commitment might inspire us. Starting with my own local context, I ask in what ways I myself exemplify the alienating modern way of living and look in detail at the place of experience in our dealings with the earth, and at the notion of living experimentally. Nietzsche's nimble, anti-moralistic but sometimes also ascetic–stoical stance towards humankind feeds into the ideas that crystallize my terrasophical approach.

In Chapter 5 I turn to the local community and its relationship with the place on earth that it occupies. It is interesting that Nietzsche attributes vital significance in his ecological philosophy to an individual's or a community's home location. He even puts that relationship with place at the heart of his concept of culture, which he sees as the mediation between people and their own particular bit of the earth. I take that thought as a starting point for my reflection on the local. I show how fruitful such a perspective can be in our own time and advocate a revaluation of local life that understands itself as connected with all other life on earth, nourishing itself with insights from an overarching planetary perspective.

In the final chapter I elaborate on the two core concepts that support the terrasophical perspective: the earth and humankind. I bring Nietzsche's insights concerning the earth and his notion of the overman into conversation with the views of contemporary philosophers, drawing in particular on the work of French sociologist, ethnologist and philosopher Bruno Latour, Belgian philosopher of science Isabelle Stengers and German philosopher of culture Peter Sloterdijk.

4

Nietzsche and 'I'

A cure for the lone individual

I dream a lot, sometimes even about Nietzsche. Not long ago I dreamed I was sitting reading *Thus Spoke Zarathustra*, studying attentively, not for the first time, his description of the meeting with the last pope. '"For know this, I am the last pope! – a festival of pious memories and divine worship." . . . Zarathustra grasped the old pope's hand and regarded it admiringly for a long time. . . . And with his gaze he penetrated the thoughts and secret thoughts of the old pope' (Z, IV 'Retired'). Zarathustra is not just able to detect hidden thoughts, he is full of compassion and alert to all secrets. What could this passage mean? Might it be an allusion to the double standards Catholicism is often accused of applying? As I was contemplating the question, drawing upon my own sentiments from a Catholic childhood, Nietzsche suddenly stormed into my room (I'd read somewhere that he did that when he was staying with friends). He looked at me with his penetrating gaze and asked, 'What is your way, actually? What is your own way?'

I woke with a start, the words still echoing, still aware of the anxiety that had taken hold of me. The dream undoubtedly had to do with a minor panic attack I experienced years ago, although on that occasion I was fully conscious. I remember exactly where it happened. I was in China. I and two colleagues had been invited to organize a conference in Beijing. An influential Chinese academic had approached us because the dialogistic method we used during conferences had impressed him greatly. That kind of conference style was unusual in China, certainly in those days (it was 2004). Only after we arrived did we hear what the conference was to be about. The subject was leadership and sustainable development. 'We would very much like to hear what a humanist has to say on the subject,' they told me, because they had read that

I was a humanist. That took me by surprise. I had not prepared anything. I had engaged with many subjects, but not that particular one. From the point of view of humanism I was unable, when put on the spot, to think of a philosopher who would be of use to me, so I stood there empty-handed, or that was how it felt. Worse still, I was deeply ashamed that the issue they rightly regarded as the key topic of the coming decades had not been a part of my humanism, my education, or even to any great extent my life. Shame and astonishment, both at my world and at myself, punched a hole in my self-satisfied professionalism.

That experience lies at the root of my own ecological revolution. In a flash it was clear to me that something I could barely name would have to become the subject at the centre of my attention in the years to come, not only professionally but socially and personally. A question that became even more important to me was: How could I have failed to see, to anticipate, this entire range of concerns? How could I have missed it so completely? In hindsight it was astonishing. There were so many signals I could have picked up and sources I could have studied. Back in the Netherlands I immediately began to educate myself, and I sought out people who could help me make further progress in the field.[1] But several years went by before I arrived at Nietzsche as the person who might be able to guide me, as a humanist, philosophically and ethically, and at first it was no more than a hunch. I therefore quite literally followed the route taken by Nietzsche in his quest, to the mountains and then to the sea, aware that the questions that concerned me most were different from those of his time.

Our own era is deeply marked by the ecological contrast experience, as I wrote in the preface to this book. We are becoming increasingly well informed about the precarious circumstances in which we live, without knowing how to take on board the insights and feelings that result. How can we find a way of life that will help us to move forward amid this fear and unease, with all its dilemmas? Which ways of thinking and acting, which life skills, do we need to acquire if we are once again to make ourselves part of life in a larger sense? And how can Nietzsche inspire and encourage us in doing so? This chapter addresses those questions. I group the ideas I have had on the subject around a number of terms that in my view point to the key elements, including those of Nietzsche's own personal quest. The first three are experience, leading an experimental life and self-transformation.

The recovery of a direct experience of nature is for modern humans the first step towards coming into physical contact with everything that lives, in order to become sensitive again. Leading an experimental life is not only about what a change in lifestyle involves, about the effort it takes, about self-denial, but also about pleasure and enthusiasm, and about the temptations of risk. Living life as an experiment is an umbrella term for adventurous lifestyle changes that lead us away from the dominant culture. Self-transformation refers to Nietzsche's way of embedding both experience and the leading of an experimental life in broader questions that people ask when they are exposed to their most profound contrast experiences and dare to allow themselves to be touched by life. Nietzsche found it necessary to have Zarathustra experience many exercises and insights from spiritual and philosophical conversion traditions. Experience, leading an experimental life and self-transformation are three separate activities, and they refer to three different skills that each require practice, or 'self-mastery and discipline of the heart' as Nietzsche puts it (HA, Preface, §4). In the interplay between them, they point to a lifestyle that I would like to call a Nietzschean art of living well, meaning one such art, not the only one, because there is no single way.

I will first expand upon these three elements and add some suggestions. After that I will look at another three aspects of the art of living well that Nietzsche sees as allied to them.

Perception and expression

I read Nietzsche's work first of all as an account of a highly personal quest to discover what it means to live on earth as an earthly creature. Anyone who wants to follow in Nietzsche's footsteps is invited to go and stand in that same arena and be spoken to and challenged by that same adventure. This requires a willingness to place in the balance one's self, one's lifestyle and one's image of humankind and the world. I now understand rather better why I first had to travel to the mountains and to the rocky bays of the Mediterranean coast. There I experienced intensely how cheerful and happy a person can feel when, after dull days, the sun suddenly comes out and throws a silver counterpane over the riotous waves. It was only as I took my daily walks along the jagged coastline and along stony cliff paths that I absorbed bodily what must have happened to Nietzsche on the rocks of Sorrento.

The accumulation of gloomy predictions about the state of the world today leads almost automatically to the idea that we must make radical changes to our lifestyles, and to a call for a new morality, but Nietzsche gives us an unexpected warning: Don't immediately start with morality lessons. For Nietzsche, in contrast to our own day, the turn towards the earth was not a reaction to a threat coming at him from the future. Instead it arose from a desire to get closer to nature. The difference is significant, because in Nietzsche it leads to the pleasure evoked by nature and to new experiences of happiness and physical well-being. There is often something graceful and merry about his turn towards the earth. The moral universe breaks open to admit an experience of nature that is as wide-ranging as it could possibly be. Zarathustra's conversion shows that the experience of value involves far more than morality alone. The new value-perspective that he has in mind revalues valuation itself. So instead of immediately giving us moral lessons, Nietzsche recommends first extending and refining the field of play of our own feelings. He initially presents not a new morality but a new sensitivity, and the development of sensory and aesthetic expression in the broadest sense. The skills needed for this are as follows: to listen attentively and look at what there is to experience outdoors and how it touches us, and to repose in nature without an aim or direction, merging with the landscape, developing empathy for animals and plants, becoming sensitive to sun, wind, rain and silence, and their effects on our behaviour and moods. This requires an active stance of receptivity and attentiveness, an alert willingness to be carried along, to submit to what happens around us and to us. By being this way in nature, day in and day out, Nietzsche was able to sense many new ideas and feelings coming to life inside him: about the sun, for instance. The sun is experienced as a generous giver, casting its light and warmth over everything that lives, every day. He muses on the idea of its generosity as the wellspring of life, and on whether this explains why we feel such delight when the sun comes out. I heal as a result of this experience; I feel more joyful, stronger and ebullient; I get an urge to dance: what does that say about what life means? Is this an effective way of expressing what is happening to me?

In humans, experiences are inseparable from expression and articulation; language and neology are therefore vital. Nietzsche developed into a true artist of language, one who aims to surprise and who, with both forgotten and new linguistic styles, awakens malevolent gaiety and offers surprising glimpses into what it means to be a bodily part of nature. He observes how rich, subtle, ambiguous, cheerful, greedy, generous, intense and callous the sound box of the body is to anyone who supplies new expressions. Even he is lost for words at that point. Nietzsche is very much opposed to what he calls 'truth pathos' (*Wahrheitspathos*), which he believed had come to dominate the philosophy, religion and science of his day. His advice is to free ourselves from this reflex and become sensitive again to the many connections between language and feeling, language and experience, language and sensation. We should then go on to ask: what do I experience more deeply, more precisely, more richly if I use different words in my contact with the earth? What language do I need to use if I want to feel that the earth is speaking to me, addressing me? Is there a language that has worked well? Which language is that?

Nietzsche became a virtuoso of language and hugely enjoyed the laughter and tears that are the bread and butter of cabaret artists. He experimented with many different genres and styles: the aphorism, the short discourse, the language of moral commands, religious or quasi-religious parables, poetry. Language is a crucial medium in the relationship between humans and nature. The choice of the form of discourse determines which experiences are possible and which words are barred. Other forms of expression are no less important. Think of Vincent van Gogh, who lived at around the same time as Nietzsche, and how sensitive he became to the effect of sunlight on his perception, always alert so that he would be outdoors at the right moment to capture the impression it made.[2] Nietzsche and Van Gogh opted to spend much of their time in the open air. The outdoor life was essential to them. Not everyone has that opportunity, especially in our own time, and the natural world around us has changed profoundly over the past century and a half. For many of us, images of nature shown in films and documentaries are more intense and impressive than contact with nature itself. Those images can of course evoke and nourish experiences of the natural world, they can awaken the desire in us to see, hear and touch it ourselves, but they can never replace the actual, direct experience. Nietzsche teaches us that we will have to find our own ways of recovering such directness, and a language that speaks to us.

Once we have acquired a richer proficiency in language and expression, with more colours on our linguistic palette, we can give scientific discourse and other forms of 'truth pathos' their due place. Nietzsche is not opposed to science. He explored the life sciences thoroughly, both biology (especially evolutionary biology) and the medical science of his time. He insisted that scientific knowledge had a special part to play in the development of the overman, and for us this is truer than ever. Scientific knowledge of the earth is even more important, complex and precise now than it was when Nietzsche was alive, although we continue to struggle with the question of how to integrate the power of this knowledge properly into a broader perspective on the relationship between humankind and the earth. I found one telling example in the work of British biologist Stephen Harding. When he went up to Oxford to study biology, it became painfully clear to him that lectures were destroying his sensitivity to nature. The scientific method forced him to adopt a stance that completely side-lined his own experience. The shock of this realization was so great that, as a means of resistance, he paid extra attention to what he perceived directly, and noted it down. He went outside as much as possible, to quiet places in the natural world, and sought new ways of amplifying and widening that sensitivity. Now a lecturer himself, he advises each of his students to seek a place in nature where they feel good, their own personal 'Gaia space',[3] and to explore all the sensory observations, emotions and thoughts that come to them there. His book *Animate Earth: Science, Intuition and Gaia* is structured in such a way that evolutionary and biological knowledge of life is interspersed with instructions for meditation on nature.[4]

Experimental life

When Nietzsche talks about experiments or experimentation, he does not mean the type of scientific testing with which we are all familiar as a way of verifying a hypothesis. Instead he always uses the term in relation to his own way of life. The terrain of the experiment is you, your comings and goings, the way you feed and clothe yourself, move and travel, work and relax, enter into relationships, live married or unmarried, like a Don Juan or like a monk, and so on. Recall what he writes under the heading 'In media vita':

> *In media vita.* – No, life has not disappointed me. On the contrary, I find it truer, more desirable and mysterious every year – ever since the day when the great liberator came to me: the idea that life could be an experiment of

the seeker for knowledge – and not a duty, not a calamity, not trickery. – And knowledge itself: let it be something else for others; for example, a bed to rest on, or the way to such a bed, or a diversion, or a form of leisure – for me it is a world of dangers and victories in which heroic feelings, too, find places to dance and play. '*Life as a means to knowledge*' – with this principle in one's heart one can live not only boldly but even gaily, and laugh gaily, too. And who knows how to laugh anyway and live well if he does not first know a good deal about war and victory? (GS, IV, §324, italics in original)

The knowledge Nietzsche is referring to is that which you need if you are going to undertake new things in your life or if, better still, a desire for innovation and creativity comes to characterize your attitude to life. Living this way is exciting and risky. It turns life into a game of dangers and victories, in which you have to be brave, prepared for failures and setbacks but in which you learn the art of putting things into perspective, of laughing and living joyfully. Could there be a better description of what might be called the art of living well?

What can persuade people to take this stance in life? For Nietzsche one of the strongest motivations was a deep dissatisfaction with the lifestyle that had shaped him and that he had made his own. He had developed a thorough loathing of the bourgeois existence in which he felt trapped, and from time to time he felt an equally thorough loathing of himself. Dissatisfaction, emptiness, disappointment and anger at the crude notions of happiness to which people had resigned themselves were no doubt important aspects, but even more so was his discovery that it was possible to see life – one's own, but also life in a general sense – as a dynamic process full of potency and ingenuity, qualities that reveal themselves in all life forms and in the case of the human species can lead to great creativity or to tragic failure. The strengths, the impulses, the passions, the capabilities that are present in every person, potentially at least, can burst out in any direction. Laws, moral precepts and religions have arisen to channel these forces. But in Nietzsche's view that time is nearly over. This is the most profound consequence of the death of god. We will now be living in an era in which we need to take that shaping in hand, starting with our own lives.

Nietzsche is interested above all in the experiential aspect of experimentation. What experiences come to people when they have an experimental attitude to life? He looks at this from a psychological, moral and aesthetic perspective and takes advice from others, from Montaigne, for example, a little of whose 'wantonness' (EH 'Why I Am so Clever, §3') he believes he has himself.

Experimentation presupposes that you adopt a stance of not yet knowing and of wanting to discover. In part, Nietzsche learns this attitude from children. Toddlers, brave and inquisitive, show us how to address life from a position of not knowing. They live in a world of perpetual dangers and victories, and every day they experience their own heroic or anxious moments, and enjoy them. For adults life is different. They have already been shaped; they already know how to live. They have settled into a lifestyle, into habits, beliefs and behaviours. Opting for an experimental attitude to life therefore demands of them that they break free from the way of life they have made their own, perhaps because it offered them security and certainty.

Experimentation and daring to let go are inseparable, as we know, but the interesting thing about Nietzsche is that he magnifies this dynamic and turns it into a heroic adventure. Because, he says, experimentation becomes truly interesting and serious only if you are about to put your life on the line, if you adopt an attitude in which the experiment concerns things that you regard as matters of life and death. How can you then remain open and unprejudiced, embracing your 'not yet knowing' and standing up to the dangers and uncertainties you confront along the way? It is important to be sensitive and courageous at one and the same time, impetuous yet extremely suspicious. Nietzsche enlarges upon the dangers and difficulties that will inevitably arise, but not in order to discourage people. On the contrary, they present a context in which people can display and develop courage and audacity, and they are necessary if we are truly to experience the happiness and pleasure that result when we set out onto the open sea, daring to leave the safety of harbour. Disruption and innovation, to use our contemporary vocabulary, presume one another.

It is this heroism that has appealed to many generations and prompted them to experiment enthusiastically, throwing off all restraint. The urge to experiment inspired the surrealists of the 1930s and the hippies of the 1960s. Philosophers including George Bataille and Michel Foucault found the courage and inspiration to subject the disciplining of the sexual urges to meticulous criticism and so to clear a path for what we have come to call the sexual revolution. But however interesting, creative and brave these innovative explorations have been, I believe they have overlooked an aspect of Nietzsche's experimental lifestyle that is extremely important in our own day, namely that Nietzsche gave a specific direction to his experimentation:

'I live so that I can discover; I want to discover so that the overman can live. We are experimenting for him' (NF-1882, 4[224]). This attitude and this motif occur several times in *Thus Spoke Zarathustra* and without this embedding, the Zarathustra experiment cannot be understood. Remember what he tells his disciples when he begins his transformation:

> Remain faithful to the earth, my brothers, with the power of your virtue! . . . Do not let it fly away from earthly things and beat against eternal walls with its wings. . . . Like me, guide the virtue that has flown away back to the earth – yes, back to the body and life: so that it may give earth its meaning, a human meaning! . . . Let your spirit and your virtue serve the meaning of the earth, my brothers: and the value of all things will be posited newly by you! Therefore you shall be fighters! Therefore you shall be creators! . . . There are a thousand paths that have never yet been walked; a thousand healths and hidden islands of life. Human being and human earth are still unexhausted and undiscovered. Wake and listen, you lonely ones! From the future come winds with secretive wingbeats; good tidings are issued to delicate ears. You lonely of today, you withdrawing ones, one day you shall be a people: from you who have chosen yourselves a chosen people shall grow – and from them the overman. (Z, I, 'On the Bestowing Virtue, §2')

It is the most dramatic passage in the book. Herein lies our task for the future. It is not a matter of an adventure that merely takes as its horizon our own life, our own happiness, beauty and consolation. The experiment Nietzsche calls on us to perform requires us to shape our lives such that they contribute to the transformation that everything will depend upon in the immediate and more distant future. By taking part in that process we will discover something that truly can be called a happy life. The heroic dynamic of breaking away and creating permeates the reasoning here too. But the experiment of one's own life is embedded in social and cultural modernity. Remaining faithful to the earth requires that we disengage from the dominant culture. Our doing so will be accompanied by experiences of being lonely and misunderstood, of having to relinquish vested interests and sacrifice prestige and standing. But – and this is the core message – do not be discouraged. There will be more and more people who choose to move in this direction. You are not alone. You will do all this along with others with whom you will share not just the low points but the successes, the progress. The undertone is of hope and comfort, of courage and confidence, despite the seriousness of the situation.

If we look at the experimental life with the overman in mind, then it becomes clear why Nietzsche makes a connection here with two specific attitudes that have received far less attention. The first is self-discipline, the 'discipline of the heart'. Experimentation relies on the ability to resist temptations that can send the passions off in any number of other directions. How can we develop the necessary toughness to enable us to ward off the influences that tempt us into other lifestyles? Who can be our guide in this respect? The other ability concerns how to discover which urges and desires we should say a wholehearted 'yes' to, which we should cherish and develop. And again: Who can help us in this respect? As regards experiments in which we take steps to the overman, therefore, Nietzsche turns the spotlight on our ability to shield ourselves from outside influences that are harmful, but above all on the art of inventing positive conduct. He formulates this as the art of learning to say a powerful 'no' and an exuberant 'yes', and then living accordingly. He sometimes refers to it as a positive asceticism, a sustained training that strengthens the will and, if it succeeds, repeatedly brings forth a new 'yes and amen song'. Zarathustra wants to be the instructor in this process, so he first tries to acquire these abilities and insights himself. Fortunately the animals are there to accompany him as he does so.

I have found many passages in which Nietzsche indicates how he intends to undertake this training himself. Here is one example:

> Not to see many things, not to hear them, not to let them approach one – first piece of ingenuity, first proof that one is no accident but a necessity. The customary word for this self-defensive instinct is *taste*. Its imperative commands, not only to say No when Yes would be a piece of 'selflessness', but also to say *No as little as possible*. To separate oneself, to depart from that to which No would be required again and again. The rationale is that defensive expenditures, be they never so small, become a rule, a habit, lead to an extraordinary and perfectly superfluous impoverishment. (EH, 'Why I Am so Clever, §8', italics in original)

Nietzsche does not want his life to be a succession of chance events. He wants to determine its direction himself and bring cohesion and coherence to it, and he regards that as a necessity, as the learning of a certain unconditionality in the way he leads his life. Being able to say 'no' is important. It means giving access only to that which assists in the quest for 'yes'. The 'no' skills are placed at the service of the 'yes' that Nietzsche is after. Because experimentation with one's own life is guided by an attitude that says 'yes' to life.

It seems a big step from this language and these images to our own situation. But is the gap really so great? If only half of what sociologists have been telling us for years about the consumer society is true – how the products we need every day and the objects with which we dress up our lives are presented by advertising, cleverly wrapped in depictions that boost our desires – are we not in our own way surrounded by the issues to which Nietzsche refers? We are surely becoming more and more firmly incorporated into strategies of consumption. Those with something to sell are using increasingly subtle and sophisticated methods to link their products to lifestyle profiling, which identifies target groups in order to appeal to their dreams and desires. No object enters our lives any longer that has not first been framed and geared to the target groups to which each of us is assigned. This approach will of course be further refined and intensified as more and more use is made of 'big data', such that advertising is adjusted according to the digital traces that every move we make leaves on the internet. Soon everything we do or decline to do will be recorded. That information is worth money because it can be used to anticipate what, based on previous behaviour, is likely to accord with our desires and needs. It will soon be possible to predict with great accuracy not only what each of us wants but what we might be persuaded to want without realizing it. The power to discover what our desires are is itself becoming embodied and part of the acceleration that will permeate the whole process. We are going to need new items, gadgets and stratagems all the time if we are to keep up. According to Bernard Stiegler[5] we are moving towards a situation in which our creative mental and emotional capacities will weaken, or even be unable to emerge at all.

Our impulsive development is itself in danger of becoming dependent on this interplay of production, consumption and technology in late-capitalist techno-society. In such a world it is surely increasingly necessary to learn how to create distance, both intellectual and emotional. If even the power to stand back a little, to offer resistance and dig in our heels where necessary, is being undermined, then surely we need to start explicitly learning and practising a new kind of self-mastery. Parents think about this when they worry about their children getting swallowed up in the world of smartphones and tablets. They can see the influence it has on their children's imagination, their use of language, their social contacts, their desires. The power of attraction is so great that we are being forced to develop new rules for interaction between children and technology, and to experiment with new educational techniques. Does all this cease to apply when childhood ends?

Nietzsche can teach us how to become more aware of the many ways in which our entrapment in patterns of behaviour and social codes comes about. He is a master at exposing the workings of social patterns. He explores the entire spectrum, from brutal oppression to the extremely subtle, almost imperceptible disciplining we undergo. He does so as a way of discovering how humans conform, how they allow themselves to be carried along and to be seduced. He also examines the fact that there have always been people who managed to put up resistance and take the helm. In his book *On the Genealogy of Morals* he reconstructs the history of morality from this point of view, and in *Beyond Good and Evil* he presents many reflections on the usefulness of accomplishments in all kinds of fields, including art and music, language and literature, and traditional workmanship, thereby providing food for thought about what a training course in life skills might look like.

Although the consumer dynamic is dominant in our time, there are many counter-movements, powered by people who want to break loose from it. One is the social development I would call increasing ecological awareness. In his book *Blessed Unrest* (2007), American author Paul Hawken looks at the ecological initiatives undertaken by individuals in America in the early years of this century. The information in the book is derived from the largest available database of non-governmental organizations active in the field. In an appendix of more than 120 pages, the thousand most important initiatives are briefly described, and Hawken explains what inspired the people involved and how readers can make contact with them. The book is intended as a contribution to an ecological World Wide Web of green citizens' initiatives, and it is subtitled *How the Largest Movement in the World Came Into Being and Why No One Saw It Coming*. This reflects precisely what stayed with me more than anything else after reading the book: no awakening has gained influence so quickly in recent years as the growing awareness of ecological imperatives. It is both a collective phenomenon and one taking place at the level of the individual. It has been compared to great emancipation movements such as the abolition of slavery, the introduction of universal suffrage, the labour movement of the nineteenth century and the twentieth-century women's movement. The expansion of capitalism means that changes in lifestyle in our own time are urgently needed. The terrain for radical experimentation in Western societies lies where contemporary consumer strategies come up against a desire for sustainable ways of living. What do people require amid this battle of skills and ideas if they are to take their lives into their own hands and discover new lifestyles? Nietzschean insights into living experimentally might give an additional stimulus to this quest.

Transforming ourselves

Nietzsche connected personal experience and personal experimentation with the question of how an individual life interlinks with life in a more general sense. Could insight into the latter help to indicate how I as a person ought to live? This issue is central to Nietzsche's reflections on his own self-transformation.

It is also the central question in ecological consciousness-raising, since it was a growing insight into the worsening state of life on earth that set off alarm bells, drawing attention to the destructive consequences for the earth of our personal and collective lifestyles. We are nowadays well informed about that dramatic deterioration. One quick way to gain an impression of where we stand as a species is to pick up the *Living Planet Reports*, in which the ecological footprint of humankind each year has been calculated as far greater than the earth can sustain in raw materials and absorb in the form of waste. In 2019 Overshoot Day, as it is known, was on 29 July, which amounts to saying that the ability of the planet to recover is exceeded annually by more than five months. If developments in production and consumption continue as they are now, Overshoot Day will continue to occur earlier each year. The footprint figure is a useful indication as to which components of our lifestyles contribute most to the destruction of the biosphere. It can be calculated for everyone's individual lifestyle, or for each country.[6] The *Living Planet Reports*, published annually, provide documented information about how quickly fish are dying, forests disappearing and sources of clean water becoming scarce. Unpolluted air is already a luxury. The worsening scarcity of clean water, energy and wood will in the foreseeable future become a source of world conflicts of an economic, political and possibly military kind.[7] It quickly becomes obvious that a change in personal lifestyle is the first step towards necessary change more generally.

Nietzsche too makes connections between the habits of individuals and the larger processes of life on earth, between the micro- and macro-perspective, between the view from below and the view from above. He gradually realizes that this will become the central issue of his philosophy. The metaphor of the pair of scales that Zarathustra takes up to weigh moral values afresh, standing on the boundary between sea and land, reflects this beautifully. Nietzsche puts the issue into words in various ways, always tentatively, experimentally and associatively. He tries through biology, physiology and the still new theory of evolution to acquire greater insight into the nature of life itself. Philosophers must become physiologists, he says. They must collect more knowledge about what the ancient Greeks called *physis* (usually translated as 'nature'). He sometimes calls philosophy a form of physics in the broad sense of the word, but he did not become a physicist, biologist or climate scientist. Nietzsche's interest in scientific knowledge mainly concerns the guidance we can gain from it for the way we live. After all, the obvious assumption is that the dynamics apparent in larger processes are at work in some way in all our lives.

For the moral aspects of this, however, he turns mainly to the history of philosophy, especially ancient philosophy and within it two schools that deal with practical wisdom, that of Epicurus and that of the Stoics. Both teach how insights into the larger interlinked systems of the cosmos provide instructions on how to live. They call it living philosophically; 'live according to nature' (*Secundum naturam vivere*) is the leitmotiv here. The Epicureans fascinated Nietzsche because their vision of nature laid the basis for a reflective, refined way of living hedonistically – not a flat, blind, impulsive hedonism that aimlessly follows impulses but a serene ability to be gladdened by life because nature invites you to be so. They taught Nietzsche an attitude of yes-saying, of sojourning in the garden of nature, and the value of a meditative style of life. But he finds their view of nature too idyllic, too Arcadian. It downplays suffering and the cruelty and inconstancy shown by living creatures. Life is full of chaos, of forces that continually battle each other. The Epicureans filter out those aspects, even though they are all too apparent in the 'sea of existence', and exclude them from their world view; in fact they project their own desire for an undisturbed, peaceful and hedonistic life onto nature. Nietzsche sums all this up in the following aphorism:

> *Epicurus* – Yes, I am proud of the fact that I experience the character of Epicurus quite differently from perhaps everybody else. Whatever I hear

or read of him, I enjoy the happiness of the afternoon of antiquity. I see his eyes gaze upon a wide, white sea, across rocks at the shore that are bathed in sunlight, while large and small animals are playing in this light, as secure and calm as the light and his eyes. Such happiness could be invented only by a man who was suffering continually. It is the happiness of eyes that have seen the sea of existence become calm, and now they can never weary of the surface and of the many hues of this tender, shuddering skin of the sea. Never before has voluptuousness been so modest.[8] (GS, I, §45)

In Nietzsche's work, Epicurus ultimately becomes a type, a mask like those of the theatres of ancient Athens, a model of idyllic life, 'of the serenity of contemplation, of the absence of strife in the cosmic vision'.[9]

Nietzsche visits the school of the Stoics because there he can learn another life skill: the art of hardening oneself as protection against temptations from outside. This is not asceticism for asceticism's sake, turning away from things as the priests do, who, according to Nietzsche, are driven by a negative attitude to life (see GM, I). In the Stoics it was a skill that flowed from a positive view of the cosmos. They developed a cosmology in which life on earth is permeated with a rationality of which the key lies in the regularity of the sun, moon and stars. In Stoic philosophy, which spans many centuries, including the period when first Greece and then Rome flourished, the idea emerged that being human means shaping your life on the basis of universal reason. Because all humans carry the 'seeds of reason' (*logoi spermatikoi*) within them, everyone is in theory able to participate in the process of reasoning that the cosmos exemplifies. It is not easy to impose this principle on our needs and desires. The Stoics saw terrestrial life, and especially the life of human passions, as inconstant. People generally seek short-lived satisfaction and are driven by their lusts, wanting to rule over others and to assert themselves. But we can give form and direction to this earthly life if we attune it to the order displayed by the universe. The Stoics thought through the consequences of taking this principle as guidance for their personal lives and their dealings with others, including those who in the prevailing culture of the time were not regarded as human in the full sense of the word (women, children, slaves and people of other cultures). The *polis* and even the entire Roman Empire ought to be guided by this principle – at least, that was the conviction of Hadrian and Marcus Aurelius, who as Roman emperors became firm adherents of Stoicism.

In the Stoic art of living well, a rational, even rationalistic approach dominates. It takes shape in a morality in which people learn to harden themselves against all emotions, needs and desires that do not conform to reason. From the Stoics Nietzsche learnt, as I have already indicated, how to harden himself for the sake of life itself. This aspect of the Stoic philosophy of life fascinated him in particular and he expresses it in his own vivid manner: 'The Stoic . . . trains himself to swallow stones and worms, slivers of glass and scorpions without nausea; he wants his stomach to become ultimately indifferent to whatever the accidents of existence might pour into it' (GS, IV, §306). But Nietzsche rejects the Stoic idea that the cosmos is an ordered reality, guided by reason. Even the Stoics failed to realize, he writes, that the most important characteristic of the cosmos is that it is without order, that unchanging laws do not exist, that life is a continual becoming, a perpetual jumble of forces and movements, which clash and battle.

As far as cosmology is concerned, Nietzsche therefore found both modern and ancient philosophers highly problematic. He did not share their view of the greater processes of life. But is a different view, a better view, to be found anywhere? Or is he forced to recognize that he cannot avoid developing a new philosophy of his own, one that takes 'chaos' and 'becoming' as its starting point and abandons the idea that life on earth has a goal or direction? That is the problem Nietzsche struggles with and the question he sets out to resolve. In my final chapter I will look at it more closely. Here I merely want to offer a long quotation from *The Gay Science*, a book written shortly before Nietzsche began his Zarathustra experiment. It shows very clearly how at the start of his transformation he wrestled with these questions. He gives the aphorism the title 'Let us beware':

> Let us beware of thinking that the world is a living being. Where should it expand? On what should it feed? How could it grow and multiply? We have some notion of the nature of the organic; and we should not reinterpret the exceedingly derivative, late, rare, accidental, that we perceive only on the crust of the earth and make of it something essential, universal, and eternal, which is what those people do who call the universe an organism. This nauseates me. Let us even beware of believing that the universe is a machine: it is certainly not constructed for one purpose, and calling it a 'machine' does it far too much honor. Let us beware of positing generally and everywhere anything as elegant as the cyclical movements of our neighboring stars; even a glance into the Milky Way raises doubts whether there are not far coarser and more contradictory movements there, as well as stars with eternally linear paths, etc. The astral order in which we live is an exception; this order

and the relative duration that depends on it have again made possible an exception of exceptions: the formation of the organic. The total character of the world, however, is in all eternity chaos – in the sense not of a lack of necessity but of a lack of order, arrangement, form, beauty, wisdom, and whatever other names there are for our aesthetic anthropomorphisms. Judged from the point of view of our reason, unsuccessful attempts are by all odds the rule, the exceptions are not the secret aim, and the whole musical box repeats eternally its tune which may never be called a melody – and ultimately even the phrase 'unsuccessful attempt' is too anthropomorphic and reproachful. But how could we reproach or praise the universe? Let us beware of attributing to it heartlessness and unreason or their opposites: it is neither perfect nor beautiful, nor noble, nor does it wish to become any of these things; it does not by any means strive to imitate man. None of our aesthetic and moral judgments apply to it. Nor does it have any instinct for self-preservation or any other instinct; and it does not observe any laws either. Let us beware of saying that there are laws in nature. There are only necessities: there is nobody who commands, nobody who obeys, nobody who trespasses. Once you know that there are no purposes, you also know that there is no accident; for it is only beside a world of purposes that the word 'accident' has meaning. Let us beware of saying that death is opposed to life. The living is merely a type of what is dead, and a very rare type. Let us beware of thinking that the world eternally creates new things. There are no eternally enduring substances; matter is as much of an error as the God of the Eleatics. But when shall we ever be done with our caution and care? When will all these shadows of God cease to darken our minds? When will we complete our de-deification of nature? When may we begin to *'naturalize'* humanity in terms of a pure, newly discovered, newly redeemed nature? (GS, III, §109, italics in original)

In his manuscript, Nietzsche underlined the word 'naturalize' (*vernatürlichen*). The aphorism is redolent with Nietzsche's struggles and despair over the ideas that present themselves as 'true' visions of life, as 'shadows of God'. But might we not make the same criticism of his view that the character of reality is chaos? Is that not also a 'shadow of God'? The final shadow, perhaps, but still a shadow. How can we think about life without getting caught in the trap of godlike thinking? Nietzsche's 'Let us beware' prepares the way for a different approach, one that does not take its guidance from the question: What is true? Instead it begins: How should we think about life, and then live it? Not long afterwards, this became the question for Zarathustra.

I believe the three fundamental requirements of a Nietzschean lifestyle are experience, the experimental life and self-transformation. Three other things flow from it that I now want to bring to the fore. The first concerns our attitude to our own past, the second our basic moral stance and the third our orientation towards the future.

A genealogical analysis of our own past

Anyone who lives experimentally and transformatively will keep bumping up against their own past. Nietzsche experienced this every day. At the start of his quest, in 1881, he made several programmatic sketches of what he would need to do in relation to his own past. They were written on scrap paper that has survived – fortunately, because although we have many pages on which Nietzsche wrote a plan of approach over the years, nowhere have I found his route so clearly and succinctly set out as in the piece to which he gave the title 'For the cure of the lone individual'. It contains what I regard as the core of his methodical self-analysis.

'For the cure of the lone individual'

1. He must start from the nearest and smallest and declare all the dependencies with which he was born and raised.
2. Equally he should understand the habitual rhythm of his thinking and feeling, his requirements for intellectual nourishment.
3. Then he should try out all kinds of changes, first breaking with his own habits (a great many dietary changes, paying meticulous attention).
4. Mentally he should get close to his adversaries; he should try their food. He should travel, in every sense. In this period he will be 'inconstant and volatile'. From time to time he should rest after his experiences and digest them.
5. After that come higher matters, the attempt to think up an ideal. This precedes something even higher: living precisely by this ideal.
6. He must pass through a whole series of ideals.[10] (NF-1881, 11[258])

This explains succinctly how to start. The first step is to take small things that are close to you and become aware of the whole chain of dependencies that have been yours all your life. So start with a concrete practice, your way of eating, for example, your way of living and relaxing. Follow your feelings of desire and unease, feelings about what does you good and what impedes you, or even makes you ill. Investigate where your ways of acting come from and expose the sequence of dependent relationships that have influenced them. This concerns your own personal past and above all what has hampered you in life. Better insight into these things is a precondition of freeing yourself from them. The energy thereby released feeds the desire to develop new ideals. This is Nietzsche's cure, which I call the genealogical-creative way of the individual. As an example, Nietzsche gives his experience with food – no accidental choice, as he explained later in *Ecce Homo*. For humans and animals, food and nutrition are the most basic and bodily connection with the earth.

What would such an approach mean in our own time? In today's food culture, supermarkets and supply chains are crucial. If we attempt to lay bare the series of dependencies that enables our food to reach our plates, we quickly find ourselves in a complex system of large-scale agricultural production, of global food imports and exports, and commercial food production that does not have 'healthy' as its main criterion. Even in Western Europe, the food environment is polluted, and unhealthy food is cheaper than healthy food.

In response there is now a trend towards the ecological and organic, especially among consumers who are free of financial constraints. Advice and information about food are improving, and people have started to pay more attention to quality. Many have found the energy to garden again, whether alone or with others, growing some of their own food. A rediscovery of gardening, of physical involvement with food-producing skills, is a way to connect ourselves with an ecological awareness that goes a long way beyond our own vegetable plots. From there we can further explore our 'dependencies' and take part in new creative initiatives, such as the Transition Towns movement.

Nietzsche did not actually pay much attention to how his food was produced. He concentrated more on the ways in which it was prepared in the German cuisine of his youth, and how that had formed and deformed his tastes and his eating habits. His starting point was the unwholesome effect German cuisine had on his body. It was one of the chains from which he wanted to free himself. Once it had been exposed and stripped of its power with appropriate ridicule, there was space for creative changes, based on a lively interest in other cuisines, including the French, Italian and Chinese. He felt an urge to experiment with his diet, to look beyond his own cuisine and make experimentation and creativity part of a higher ideal, and then to live by it. In the early 1880s that ideal came to be called the overman and Zarathustra became the figure who would demonstrate to the full with his own life the 'cure of the lone individual'. Eating and fasting are given a broader significance as steps to the overman. The animals are not stock. They point the way, and are delirious with joy when Zarathustra moves in the right direction.

We can further expand Nietzsche's analysis of food not just by asking what we are eating and how it is prepared but by reflecting on the question of 'how?', here too exposing dependencies. Our culinary culture has changed profoundly, and food can now be bought ready to eat. Do we eat alone or together, in front of the television or at a table, à la carte or out of a common pot, with our hands or with cutlery? Each individual can explore the options and contemplate whether current ways of eating are doing them good, whether they are becoming stronger and more vital as a result, and whether there is some kind of ritual involved that connects food with a higher ideal.

Regarding that ideal, let me reveal some of my own dependencies here. I was brought up in a time when grace was said before meals. In my childhood we were not allowed to start eating before my mother had said, 'God, we thank you for the gifts that we are about to receive from thy bounty...' At my Catholic boarding school, we could not 'attack' the food before the housemaster had said a prayer, even though the ritual had become a formality. In the refectory, where three times a day food was served to several hundred boys aged between thirteen and eighteen, strategic positions were adopted during the saying of grace, with an eye to the serving dishes. In the years that followed I was a monk for a while, a Franciscan. Monastic life introduced me to a rhythm of singing and eating and fasting. I have many positive memories of that time, but I also witnessed forms of asceticism that brought nobody any joy. After that I became a student again and experienced at first hand the cultural revolution of 1960s Paris. How exciting and liberating it was to set out to smash all those chains of dependency. Now I sometimes find myself thinking that perhaps it is time once more for a bit of ritual, for a framework to remind us that there is a vital, bestowing link between the earth and food. Time for a simpler lifestyle too, perhaps, designed to remind us that the earth's resources are not inexhaustible. Nietzsche's reorientation towards the earth went along with a sober but nonetheless joyful lifestyle. Such was his life as he chose to portray it in his writing.

Nietzsche sought the genealogical route of self-analysis not just in relation to food but for many other subjects drawn from his life, and he did so far less bookishly than I have described here. His tone is more playful, less emphatic, more surprising. The cure for his sick body and the search for 'the great health' are inspiring material for anyone wishing to follow Nietzsche in this purview of life. Barely a day went by when he did not suffer from migraine, and his eyes were so bad that he sometimes had to have others read to him, or dictate his work to his sister, a friend, or a willing companion. His weak and ill body took him back to his family past, to what had been handed down to him of his father's physical and mental afflictions, a father who died when Nietzsche was just five years old. That loss preoccupied him all his life in his dealings with sickness and death, and with religion. Another genealogical line can be followed concerning his dealings with women, with Malwida von Meysenbug and the many others who liked to be in his company. We see how the beliefs of his time affected him and witness the impact of the most painful episode in his love life, his rejection by Lou Andreas Salomé. Again and again his guiding thought was: How do I liberate myself and get over this? And for what purpose? For which values? In Nietzsche's exploration of the relationship between humans and the earth, his relationships with women may be just as important as his relationship with diet and health, but they would require another book.

A yes-saying morality

> And all in all and on the whole: some day I wish to be only a Yes-sayer! (GS, IV, §276)

When it came to morality, Nietzsche sought a positive attitude to life. He formulated his own moral compass as follows:

> At bottom I abhor all those moralities which say: 'Do not do this! Renounce! Overcome yourself!' But I am well disposed toward those moralities which goad me to do something and do it again, from morning till evening, and then to dream of it at night, and to think of nothing except doing this *well*, as well as *I* alone can do it. When one lives like that, one thing after another that simply does not belong to such a life drops off. Without hatred or aversion one sees this take its leave today and that tomorrow, like yellow leaves that

any slight stirring of the air takes off a tree. He may not even notice that it takes its leave; for his eye is riveted to his goal – forward, not sideward, backward, downward. What we do should determine what we forego; by doing we forego – that is how I like it, that is my *placitum*. But I do not wish to strive with open eyes for my own impoverishment; I do not like negative virtues – virtues whose very essence it is to negate and deny oneself something. (GS, IV, §304, italics in original)

The terrasophical approach is not moralistic in the familiar sense of the word. Nietzsche does not admonish based on a moral consciousness, let alone a moral truth. The potential for change does not arise from knowing better, from a commandment or a corrective. It is not '*ecologie punitive*' (punitive ecology), to quote former French environment minister Ségolène Royal. In Nietzsche the impulse for change has its source in a longing for 'the great health'. A whole range of new initiatives may arise from that one impulse. Creativity and plurality go hand in hand because the stillness of the earth invites free expression and ambiguity. Self-mockery, humour and the liberating effect of irony counteract the moralizing raised finger.

Nietzsche was not the only person to connect the art of personal engagement with the art of putting oneself into perspective; Kierkegaard too, at around the same time, gave good examples of this. But the true master was for me Mahatma Gandhi, who might be called the Indian Zarathustra. His approach has a surprising amount in common with Nietzsche's. Gandhi's most influential book is his diary, published as *An Autobiography: The Story of My Experiments with Truth* (1927), in which he recorded how he brought himself to greater authenticity and how important in doing so was his regime of vegetarianism and fasting. For Gandhi a personal lifestyle was the route to a radical liberation from the chains of colonial domination, which he experienced as the main obstacle to his future and that of India. With his hunger strikes and his mass marches and walking protests in his famous white dhoti and shawl, he ultimately brought the British rulers to their knees like a true Zarathustra. He advocated a politics that had at its heart an ecological lifestyle derived from Indian culture. When Gandhi was asked for guidance, he said, 'Be the change you want to see in the world!'[11] And when in London a journalist asked him for his opinion on Western civilization, he responded with memorable wit: 'I think that would be a very good idea.'

Scenarios for the future

As well as a richer experience of nature, an experimental attitude to life, self-transformation as a motive, a genealogical-creative self-analysis and a yes-saying moral stance, I want to bring to the fore one last element of Nietzsche's personal approach that can be an inspiration for our own time. His entire art of living well is radically focused on the future.[12] Just how essential this orientation towards the future was for Nietzsche can best be made clear with the help of quotations from *Thus Spoke Zarathustra*.

Zarathustra asks the disciples who have followed him, 'This – it turns out – is *my* way – where is yours? ... *The* way after all – it does not exist!' (Z, III, 'On the Spirit of Gravity, §2', italics in original) There then follows an extraordinary scene, entitled 'Of Old and New Tablets', a reference to the two stone tablets Moses brought down from Mount Sinai, given to him by Yahweh, with the Ten Commandments engraved on them. There are no commandments on these new tablets, only hints, powerfully formulated, and suggestions for the future. We might call them 'attempts' or 'endeavours'. There are thirty and, as we might expect, they concern all the important dimensions of existence. They run through all the instructions designed to connect our personal lives with the fate of generations to come. '*There* our helm wants to steer, where our *children's land* is! Out there, stormier than the sea, storms our great longing!' (Z, III, 'Old and New Tablets, §28', italics in original) Change in the lone individual therefore has reasons that do not lie merely in the individual concerned, as if the horizon of our happiness stretched no further than our own self-preservation. 'You should love your *children's land*; let this love be your new nobility' (Z, III, 'Old and New Tablets, §12', italics in original). Breaking free from one's own genealogical past features here too: 'You should *make it up* in your children that you are the children of your fathers; *thus* you should redeem all that is past! This new tablet I place above you!' (Z, III, 'Old and New Tablets, §12', italics in original) I read 'fathers' as 'colonial fathers', a point I will come back to in the next chapter. Seen from the perspective of the earth and the overman, the life of the individual is interwoven with the threads of previous generations, and above all with generations yet to come.

It took a long time for this intergenerational idea to be formulated in contemporary discussions of ecology. The breakthrough came in 1987 when the report *Our Common Future* by the UN's Brundtland Commission came up with the following definition: 'Sustainable development is development that meets the needs of the present without compromising the ability of future generations to meet their own needs.'[13] Since then, intergenerational justice has expanded to become a new category of ethics.[14] Nietzsche points to a different aspect. He is not merely concerned about a fairer distribution of necessities between generations, as if each has a right to a proportionate slice of the cake. He describes a different vision of life. The modern idea that people are independent individuals has led to the severing of the connection with generations before and after. From the perspective of the overman, people will rediscover the bond with other generations as an essential dimension of a 'grounded' way of life. Modern individualism produces isolation, Nietzsche believes and narrow ideas about what makes people happy. He advises his contemporaries, who overestimate themselves, gradually to divest themselves of the 'austere shirt of duty' in which the individualism of modern culture constricts them and to develop a future in which people will rediscover their kinship with other generations, and with the animals, indeed with all forms of life on earth.

Whether or not life on our planet has a purpose was a question that preoccupied Nietzsche intensely. He was critical of all philosophies that claim to be able to discover purposefulness in the emergence of life and especially in human history. He criticized on this point the great teleological thinkers of the nineteenth century, Hegel and Marx, and those theorists who regarded evolution as purposeful. His discovery that life demonstrates eternal recurrence convinced him in his opposition to teleology, but that left open the question of whether the life of the individual was without direction and purpose. How are we to understand this in relation to Zarathustra's passionate call to us as modern people to change and take 'the steps to the overman'? How are we to understand Zarathustra's assertions that today's human beings are a bridge and not a purpose, a transitional phase in the development of humankind, a phase that has to be overcome, that the great thing about humans is that they are both downfall and transition? Here we come up against an important distinction between teleology and meaning. Teleology investigates whether there is a goal inherent in life that steers evolution from within, whereas meaning concerns whether we create a purpose for ourselves, a direction, and what that signifies. To this second question, Nietzsche gives an explicit answer in *Thus Spoke Zarathustra*. The meaning of our lives lies in a new relationship with the earth, a way of living that is in harmony with nature. How can we rediscover ourselves as part of nature? It is a question that indicates a direction for the future. Nietzsche cannot tell us what the road ahead will look like. It is up to us and every new generation to discover that for ourselves. What is certain is that we moderns still have a long way to go. We are 'a bridge and not an end'.

5

The place on earth

A revaluation of the local perspective

> *Cultural diversity is as necessary for humankind as biodiversity is for nature. In this sense, it is the common heritage of humanity and should be recognized and affirmed for the benefit of present and future generations.*
> UNESCO Universal Declaration on Cultural Diversity (2001)[1]

Nietzsche saw the earth, inhabited by humans, as a planet that would continually be changed by human activity. The notion of pure, virgin nature was alien to him. This outlook lay at the root of his metaphor of the garden, which he consistently used to describe a permanent transformation of the earth. The metaphor needs to be understood in a broad sense. If we put together all Nietzsche's references to gardens and the happiness of gardeners, they evoke 'a variety of activities, dispositions and tastes. . . . The dominant themes are the shaping and tending of the natural, with a view to producing a rewarding result as well as the enjoyment of the Earthly site.'[2] References to gardening can be interpreted as references to the cultivation of nature, in the oldest sense of the word 'cultivation'. Nietzsche often refers to culture, a word derived from the Latin verb *colere*, which in its original meaning unites 'to farm' and 'to foster', covering all the implications of his garden metaphors: to cultivate, till the soil, inhabit, venerate and embellish. Gardening does not refer only to individuals working the soil, whether out in the world or inside themselves as described in Chapter 2. Nietzsche uses the metaphor mainly in a general sense, to make visual the relationship between humans and their place on earth. It has acquired an additional connotation in today's world because far from gardening, the urbanization of the earth is now the apparently unstoppable trend, and there are more and more megacities with over a million inhabitants. Cities have become places where people are cut off from the natural environment, losing all contact with nature. We are beginning to realize how much damage is being done by this way of inhabiting the earth. Nietzsche could know nothing of it, yet in his own time he wondered what cities might need to look like in order to remain habitable:

> One day, and probably soon, we need some recognition of what above all is lacking in our big cities: quiet and wide, expansive places for reflection. Places with long, high-ceilinged cloisters for bad or all too sunny weather where no shouting or noise of carriages can reach and where good manners would prohibit even priests from praying aloud – buildings and sites that would altogether give expression to the sublimity of thoughtfulness and of stepping aside. (GS, IV, §280)

Large-scale urbanization is starting to alarm us. We are increasingly aware that we must not allow the way that people live to damage the earth. Nietzsche's call to cultivate the earth as a garden makes us all the more sharply conscious of our responsibility as inhabitants of a living planet. It is a theme I want to pick up from a genealogical perspective, in order to investigate how, over the centuries, communities have adapted the natural world around them. The main questions I address in this chapter are as follows: How does the patch of earth that a group of people has settled into – be they an extended family, a village community, or an entire nation – come to life in their culture? How does it take shape in their habits and customs, their social mores, rituals and art? Most importantly: What can we learn from this for the future? How can we develop a worldwide 'gardening culture' that gives meaning back to the earth and hope back to humankind? In his essay 'The Wanderer and His Shadow' Nietzsche dreams that the earth will one day be a 'collection of health resorts', where people can experience all the dimensions of their gardening happiness (WS, §188, as quoted in Shapiro, 2016, p. 160).

Shapiro devotes an entire chapter in *Nietzsche's Earth* to the subject, under the heading '"The World Awaits You as a Garden": A Political Aesthetic of the Anthropocene' (2016, pp. 134–165). It includes an extensive description of eighteenth- and nineteenth-century landscape architecture, supplemented by detailed accounts of the significance of landscapes in the paintings of the time. Nietzsche may have been influenced in part by these traditions, Shapiro concludes.[3] He claims that Nietzsche's garden metaphors are attempts to sketch out an aesthetic politics of the earth. Shapiro goes on to develop the idea[4] that Nietzsche was seeking an aesthetic that was 'consistent with his general philosophy of art, which stresses the power of framing, shaping, and making' (2016, p. 156).

I want to make a connection between Nietzsche's metaphor of the garden and a line of thinking other than the aesthetic, namely that of ethnology, the scientific discipline that since the late nineteenth century has attempted to describe and interpret local cultures from a non-Western perspective. Nietzsche was not familiar with the discipline as we know it today, but the idea of looking at modern culture through the lens of non-Western cultures would certainly have appealed to him. The ethnological approach is already present in *The Genealogy of Morals*, to which, as Deleuze writes in *Nietzsche and Philosophy*, 'all modern ethnology owes an inexhaustible "debt"'. The studies I have consulted operate on the interface between ethnology and ecology. The term 'biocultural' has been introduced to refer to the intrinsic connection between biological and cultural aspects of a society. A 'biocultural region' is 'a local geographic area in which specific human cultures develop in relation to the natural ecosystems they inhabit'.[5] With the help of information from biocultural research I try to answer the question: How have communities, starting with indigenous cultures, embedded themselves in their home regions and how has this changed them and the land they occupy? Above all, what can we learn from this for the future? One might call it the start of a genealogy of gardening in its broadest sense.

I have chosen this perspective because it helps me to bring out the potential of Nietzsche's thinking about turning the earth into a garden. In Chapter 2 I indicated that in Nietzsche there are two lines of thinking about bonds with the local environment. On the one hand there is his radical opposition to the rise of nationalism and fascism, which integrate local customs and traditions into an aggressive state politics. His aversion to politics of this kind led Nietzsche to approve of the idea of making himself into a good European. But another line of thinking becomes visible in the way in which Nietzsche pays attention to the bodily experiences of living, gardening and feeding oneself, and to the local habits and mores in which these are embedded, often referred to as *Kultur*. Aware of the fact that he needed to free himself, bit by bit, from the 'austere shirt' of abstract and spiritualized concepts that constricts modern humans, Nietzsche decided to philosophize in a way that adheres to the 'guide of the body'. He stays close to everyday experiences and practices, which are the terrain of ethnology. He advises us to step outside our own culture as part of a critical analysis of culture as a whole. Only by looking at our own culture through the eyes of other cultures are we able to see things that we miss from the insider perspective.[6]

This chapter is structured as follows. I start with two ethnological studies of indigenous cultures, which present a very different vision of the cultivation of the earth from that of the modern West. Today's ecological problems have brought the ways of life of indigenous peoples to our attention afresh. A great variety of native cultures still exists. The findings of these two ethnological studies are then supplemented with examples from contemporary local ecological awareness in countries commonly referred to as 'emerging economies'. They illustrate how in situations other than those of the West, an interaction between ecology and local awakening can take shape. Against the background of this global view, I then return to the situation in the West and look at the values modern culture attributes to the relationship between the community and its specific place on earth. This raises two questions for further research. The chapter ends with a philosophical reflection on 'becoming indigenous'. The detailed journey through other cultures in this chapter is, as I have said, not one that I have undertaken on my own initiative. I am following Nietzsche's firm advice to look at our own culture through the eyes of other cultures.

Local culture: An ethnological perspective

In 2000 the World Wildlife Fund (WWF), in collaboration with Terra Lingua, published a large-scale study of the relationship between biodiversity and cultural diversity. It took an innovative approach.[7] On a map of the world, it marked the ecological regions that were of particular importance for the preservation of biodiversity on earth. These included the tropical rainforests, tundra, coral reefs and mangrove swamps. Of the 895 regions that fitted the category, more than 200 were chosen as being of 'exceptional importance'. They have since become known as the Global 200. It proved a successful strategy for highlighting those regions' vital importance for the preservation of diversity worldwide. Attempts were then made to identify the human communities that had helped to conserve them, and as a result it became clear that in regions of great importance for the preservation of biodiversity, many native peoples lived or had lived, with cultures that were very much their own. They came to be called 'ethno–linguistic groups', because the researchers believed that having a language in common is an important element of a culture. The study reached a conclusion the local inhabitants had themselves been trying to communicate for centuries, namely that the preservation of biodiversity is inextricably bound up with the activities of indigenous peoples. In other words, it is no accident that areas where the original inhabitants still live are areas where biodiversity is protected, as long as the humans have managed to continue to adhere to their own cultural values.

The results produced by the WWF and Terra Lingua research are supported by many ethnological studies. Ethnologists, linguists and biologists had already shown repeatedly that indigenous cultures and local communities possess detailed and sophisticated knowledge of the flora and fauna in their regions, and on that basis have developed ways of living that not only contribute to the preservation of biodiversity but, according to Pascal Picq, actually manage to increase it.[8] Their knowledge and ways of life cannot be seen in isolation from their natural environment. Dealings with nature are expressed by indigenous peoples in relational terms. Animals and plants, forests and lakes are to them partners with which they maintain all kinds of connections, including kinship ties. The results of these studies are of great importance for the analysis of modern consciousness. They present a mirror image of the consequences that the loss of such local cultural consciousness has for biodiversity. Clearly the ecological damage that modernization brings with it has robbed us of valuable players in the process of transforming the modern human into the overman.[9]

An important place in ethnology is reserved for the work of French professor Philippe Descola. Following in the footsteps of Claude Levi Strauss, who produced ground-breaking work in the field in the twentieth century (Picq, 2013), Descola carried out a comparative study of all cultures, whether indigenous or modern. He compared the ways in which they each shape the relationship between humans and other living creatures, especially animals and plants, which he calls 'non-humans'. His research asks: What is the relationship between humans and non-humans in existing cultures? Descola was led to the conclusion that in the great variety of cultures that can still be found on earth, we see four basic patterns. He calls them naturalism, animism, totemism and analogism. These are four different world views, which he also refers to as cosmologies, ontological frameworks or ontological routes.

Naturalism is the pattern of modern culture that began with the Renaissance. Its characteristics are now reasonably well known. It draws a radical dividing line between humans and non-humans, attributing a lower status to animals and plants, which are seen as having lower forms of consciousness and no inner life at all, in contrast to human beings, whose unique status is given to them by their rationality and autonomy. This creates a dichotomy in experience and esteem. Non-humans belong exclusively to the domain of nature, which not only is separate from that of humankind but stands in opposition to it. They are mere objects. It is only a short step from there to the objectification of nature as a whole, of which objective knowledge is possible. Descola writes that the objectification of nature is not a result of scientific knowledge. Rather the scientific and objectifying approach had first to be made possible by the drawing of a line between 'human' and 'non-human' as a template of the culture.[10] The fact that Descola refers to modern cosmology as naturalism may seem surprising – I would rather have expected the term 'anthropocentrism' – but he favoured the word 'naturalism' as a way of emphasizing the split between humans and non-humans and the object-status given to nature.

What makes Descola's comparative study so interesting is that he situates the Western cultural pattern in the broader context of the other cultures that still exist in the world. From that comparison modern culture emerges as the odd one out, because on the subject of the relationship between humans and non-humans, animism, totemism and analogism propagate very different ideas. All three share a number of fundamental characteristics that make them radically different from naturalism. First, they all recognize that there are fundamental similarities between humans, animals and plants, and that all three are related and interdependent. At the root of these cultures lies the belief that all living creatures are part of a single collective. The relationships between them are those of kinship, reciprocity and mutual dependency. Secondly, all three cultural patterns are characterized by a firm solidarity with the earth, which these cultures regard as the largest life-giving entity. The local environment is pervaded by 'natural forces' that transcend that particular locality, forces that are named, portrayed and celebrated by the culture. Thirdly, all three are in a historical sense far older than modern naturalism. They have roots that go back more than a thousand years, and in the case of the totem culture of the Australian Aboriginals, for instance, far further still.

In the animist culture found among peoples of the Amazon basin, parts of North America, Siberia and some regions of southern Asia and Malaysia, an inner life is attributed to all living creatures. Living beings are spiritual beings; in that respect humans, animals and plants do not differ. Between the living, and even between the living and the dead (ancestors who live on as 'spirit animals'), there is therefore a spirited and inspiring exchange. The difference between them does not lie in their inner consciousness but in their physical exterior, their manifestation. This means that living together is not always easy – because a human must not simply hunt and kill animals, for example. The hunt is accompanied by rituals intended to reassure the animals or explain to them why it is taking place, since otherwise they might avenge themselves or disappear. Animistic cultures surround the delicate balance within the collective with many rules, customs and stories.

In totem cultures, of which those of the Australian Aboriginals are the most important, stories have perhaps an even greater part to play. These cultures describe how animals and plants occupied the land before humans arrived and possess greater wisdom concerning life in the region. Humans can learn from them. Every family within a given people has a relationship with a specific place, plant and tree, and with a specific animal that is its totem. This relationship means, among other things, that a family and its animal adopt each other and share specific qualities of the totem.

Analogism is characteristic of cultures of Chinese origin. Typically, all elements within an analogist culture are ontologically separate but connected by analogies or correlations that create an organic and dynamic whole. One well-known example is the correlation between the body and the sole of the foot, which communicates with all the internal organs. Analogies may also be found between humans and the firmament. The following classic text presents a good example: 'Heaven has four seasons, five phases, nine sections, and 366 days. Man likewise has four limbs, five viscera, nine orifices, and 366 joints.'[11] The use of 'likewise' in this way is common in Chinese philosophy. It has a power of persuasion that it has lost in Western thought.

Descola examines in detail the different cosmologies and the variations within them, making them accessible to our Western way of thinking. One of his aims in doing so is to show that the depiction of the relationships between humans, animals and plants that we take for granted, based as it is mainly on usefulness, is just as much a cultural construct, one that, in comparison to others, has caused a good deal of ecological damage. From an ethnological perspective, therefore, modern naturalism is a latecomer and an outsider among human cultures. The title of the hefty tome in which Descola sets all this forth is *Beyond Nature and Culture* (Chicago 2013), or in the original French *Par-delà nature et culture* (Paris 2005), a title that clearly contains a reference to *Beyond Good and Evil*, one of the books in which Nietzsche criticizes modern culture.[12]

An ethnological view of European history

Ethnological studies enable us to look at the European past in a new way. The great colonial powers of Europe that conquered the world from the sixteenth century onwards, and whose dominion ran in parallel with the development of modernity, attributed little if any value to native cultures and their ecological knowledge and wisdom. Colonial strategy consisted above all of stripping the local peoples of their supposedly primitive consciousness of the earth and their existing religions, an approach justified by Christianity and by the Enlightenment concept of progress. Even today, many people think that indigenous cultures are destined to disappear in the ongoing violence of cultural and economic globalization. 'Most well-informed people assumed that genocide (tragic) and acculturation (inevitable) would do history's work.'[13] But is it that simple? Surely something very different arose in the late twentieth century. New forms of 'native wisdom' seem to be emerging again everywhere and involving themselves in the battle for a new consciousness of the earth. The cultural tenacity and struggles of indigenous peoples confront us with the fact that our own culture no longer makes us feel at home in the place on earth where we live. In modern culture, the relationship between humans and their natural environment disappeared as a subject worthy of attention. People became alienated from their local roots.

In the research field that takes the relationship between community, culture and environment as its subject, a place of honour should be given to indigenous cultures in the strict sense of the term. After all, they have never ceased resisting the destructive side effects of modern, de-localized culture, despite the high price they have paid for such resistance. 'Natives' have been branded primitive. They have been humiliated, enslaved and tortured. They were put on display as a species of ape in the European capitals of the nineteenth century, converted or forcibly baptized, and drugged. In this history of suffering, huge numbers were killed, but among the survivors, resistance ultimately triumphed.[14] We might say, with Nietzsche, that 'great pain, the long, slow pain that takes its time – on which we are burned, as it were, with green wood' (GS, Preface, §3) has made them all the more tenacious and has deepened both their scepticism and their insight. Nietzsche looks to the indigenous person for lessons on how to face such suffering head raised, with pride, willpower and scorn. We should do so by 'equaling the American Indian who, however tortured, repays his torturer with the malice of his tongue' (GS, Preface, §3). Perhaps our attitude to indigenous peoples ought to follow Zarathustra's advice: 'You should *make it up* in your children that you are the children of your fathers; *thus* you should redeem all that is past! This new tablet I place before you!' (Z, III, 'On Old and New Tablets, §12', italics in original). Based on what has gone before, I suspect that by 'fathers' he means colonial rulers.

Experts, in the field of development aid for example, sometimes warn against idealizing or romanticizing indigenous cultures. They claim that in any case native peoples are now inextricably caught up in globalization and that in their pure form, indigenous cultures no longer exist, with a few extraordinary exceptions. It is certainly true that, in our own time, indigenous peoples are rarely untouched by today's world. Furthermore, they seek each other out, creating international platforms and making use of mobile means of communication and social networks. Their 'cultural rights', at last recognized by the United Nations, are effectively interwoven with their struggle, so that they now seek media attention and sometimes succeed in marketing their natural products. Globalization and decolonization have provided them with new opportunities for mounting resistance, but it remains the case that those who continue to defend their indigenous cultures are generally poor and lack adequate healthcare and suitable educational opportunities. Many have joined cultural movements such as those of the anti-globalists. The problems that have come to the fore are different from those of fifty years ago, and other opportunities for change are being explored. 'Cultural endurance is a process of becoming,' ethnology professor Clifford is right to claim (2013, p. 7). Their current situation raises new questions about the future of indigenous cultures, and to those questions Clifford, who knows the terrain better than anyone, answers,

> Indigenous peoples have emerged from history's blind spot. No longer pathetic victims or noble messengers from lost worlds, they are visible actors in local, national, and global arenas. On every continent, survivors of colonial invasions and forced assimilation renew their cultural heritage and reconnect with lost lands. They struggle within dominant regimes that continue to belittle and misunderstand them, their very survival a form of resistance. To take seriously the current resurgence of native, tribal, or aboriginal societies we need to avoid both romantic celebration and knowing critique. An attitude of openness is required, a way of engaging with complex historical transformations and intersecting paths in the contemporary world. I call this attitude realism. (2013, p. 13)

He even goes a step further, convincingly defending the assertion that 'becoming indigenous' is developing into a 'historical force', a narrative of its own alongside globalization and decolonization. Becoming indigenous is the third narrative that we need if we want to understand what is going on in the world today.[15]

Drawing upon the perspective of indigenous cultures

I want to stress, within the framework of terrasophy, two central assumptions shared by all indigenous cultures that directly affect the relationship between the local and the ecological.

The first concerns their cosmology. Indigenous cultures model their relationship with the environment on partnership and kinship. They experience themselves as part of the intricate web of life and honour the earth as the mythical Gaia, a mother who feeds and cares, and is deserving of gratitude and respect in return. They bring back to life the symbols, myths and stories, the works of art and management practices appropriate to this attitude, along with the notion of an 'integrated culture', one that closely links its spiritual, moral and symbolic dimensions with practical, everyday life, with upbringing and education. This culturalization of one's own environment retains its historical and symbolic value in the present context. Gaia cosmology can generate new energy and creativity in our own time and contribute to a new global ecological culture.[16]

Closely bound up with this is a circular experience of time, as clearly explained in the following quote from Epeli Hau'ofa of New Guinea.[17]

> Where time is circular, it does not exist independently of the natural surroundings and society. It is important for our historical reconstructions to know that the Oceanian emphasis on circular time is tied to the regularity of seasons marked by natural phenomena such as cyclical appearances of certain flowers, birds and marine creatures, shedding of certain leaves, phases of the moon, changes in prevailing winds, and weather patterns, which themselves mark the commencement of and set the course for cycles of human activity such as those related to agriculture, terrestrial and marine foraging, trade and exchange, and voyaging, all with their associated rituals, ceremonies, and festivities. This is a universal phenomenon stressed variously by different cultures.[18]

It is a phenomenon all indigenous cultures have in common. I highlight it here because Nietzsche seems to assume that it will be part of a future ecological culture. He connects a circular experience of time with a life lived in connection with the earth.[19] He then contrasts it with modernity's linear and teleological experience of time, which introduces a perspective based on the concept of progress. The tension between circular and linear experiences of time strikes me as a promising topic of conversation between the two cosmologies.

Local ecological awareness in emerging economies

Indigenous peoples represent small communities with a clear identity. In addition, however, in Latin America, Africa, India, China and Indonesia a large proportion of people still live in what are known as rural agricultural settings, even though many are now leaving for the cities.[20] In rural areas, interaction between ecology and local environmental consciousness is increasing rapidly. In India, Africa and Latin America it is inseparable from the deep disappointment that arose in the years after national independence. Former colonies quickly transformed themselves into modern states, each of which wrote its own constitution. In no time the modern form of the state and the discourse of modernization that went with it became the dominant political model. Within a mere eighty years or so, the surface of the earth was divided into territories, each with its own rulers. The state, which

Nietzsche describes as 'the coldest of all cold monsters', seized power. Now that the political model of the modern state has been established everywhere, its downsides are clearly visible. Political elites have arisen that strengthen themselves militarily as they see fit and enter into profitable deals with multinationals, with something known as 'development' as their aim. Where the alliance of globalization and decolonization shows its face, the faith in political leaders that existed initially (think of Gandhi and Nehru, of Mao and Che Guevara and, more recently, Nelson Mandela) is quickly destroyed. The process that Clifford observes in indigenous peoples in the strict sense is manifesting itself far more broadly. At very many places on earth, 'natives' are emerging from a blind spot in history. Everywhere they are seeking new ways of taking back the land they live on.[21]

Although such initiatives are still fragmentary and do not have much clout on a global scale, their potential is greater than ever. A wide range of strategies is deployed, from radical confrontation with the prevailing state power to the setting up of joint projects in collaboration with that same state power. In parts of the world such as India and Latin America that have a long tradition of political and cultural revolutionary resistance, this energy is now being mobilized as part of the increasing trend towards ecological awareness. One telling example is the movement initiated in India by Vandana Shiva. She draws a firm distinction between the intertwined interests of the state and globalization on the one hand and on the other the building up of local communities, with their local knowledge, local culture and craftsmanship. On that basis, links are then made with the latest discoveries in sustainable technology, making use of all other technologies that prove their worth. Vandana Shiva trained as an engineer and is a respected academic. She is now applying her expertise to this new strategy, and many other scholars are exploring similar ideas.[22] In Latin America the *buen vivir* movement adopted the same approach, and increasing endorsement of it by the population has led to the incorporation of its ideas into new constitutions in Ecuador (2008) and Bolivia (2009).[23] An example of collaboration between the state and local communities can be found in Uganda, where national politicians and local populations have come together in an effort to align state legislation and customary law.[24] In short, in the politically independent countries that until a few years ago were still known as 'the third world' potentially valuable local ecological initiatives are flourishing, although we might well wonder whether they will be able to develop into a powerful force in places where globalization has a considerable head start.[25] This trend is supported philosophically and academically by studies on the subject of bioregionalism,[26] a model that favours small, local communities linked in a network of relationships, with core values of equality, sharing and the creation of an ecological balance between people and their environment. The aim is to create societies that are self-supporting and self-regulating. The extent of each community is that of a natural region and any local and ecological culture that already exists there.[27]

The issue of urbanization

These local rural initiatives defy the tendency to move to the cities, which continues apace in emerging economies and seems irreversible. All across the world, people have been migrating over the past sixty years to cities that are increasingly big and complex. Estimates suggest that 3.5 billion people now live in cities, or a little over 50 per cent of the world's population. By 2050 this will have risen to 75 per cent, or 6.2 billion, inevitably leading to a growth of existing cities. Whereas Germany, for example, has 3 or 4 cities with more than a million inhabitants, India now has 46 and China 160. Urbanization is a central feature of upcoming economies, and most of these cities are producing toxic environments. The Global Footprint Network concludes that worldwide efforts to create sustainable societies will be won or lost in the cities, since urban development accounts for more than 70 per cent of the global footprint. Already, 80 per cent of CO_2 emissions are the result of urban lifestyles.[28] It is important to realize that the dominant human type of the twenty-first century may well be a resident of one of these super-cities, born and brought up in what is already a heavily polluted urban environment. People who live in big cities are far from the natural world, and the basic necessities of life (food, housing, transport, communications) are dependent on complex social and economic chains of cause and effect whose workings are largely hidden.

Terrasophy cannot escape the need to rethink the relationship between the city and its surroundings from the perspective of the earth. Nietzsche offers us little help in this respect. He uses 'the city' as a metaphor for a way of life that alienates people from their connection with the earth, but this obstructs our view both of the historical role cities have played in human development and of our current situation. Were cities in the past not centres of innovation and creativity? In his book *Triumph of the City*, Edward Glaeser writes, 'The central insight is that proximity makes people more inventive, as bright minds feed off one another; more productive as scale gives rise to finer degrees of specialization; and kinder to the planet, as city-dwellers are more likely to go by foot, bus or train than the car-slaves of suburbia and the sticks.'[29] John Elkington, who quotes Glaeser in his book *The Zeronauts*,[30] points to further positive sides of urbanization, for the position of women, for example. 'Urban women tend to have fewer children, partly because they have better access to family planning, and partly because they generally have better employment prospects, too. "Women in cities stay in school longer, have better access to contraceptives, get married later and have their first child later," explained Jocelyn Finlay, based at the Harvard Center of Population and Development Studies' (2012, p. 93). In other words, no one-to-one relationship exists between urbanization and poorer prospects in life, far from it, but it remains the case that today's rapid urbanization has side effects, especially in developing economies, of which the impact in ecological terms has not been fully calculated by any means.

The West and the local

I have paid a good deal of attention to the subject of location in other parts of the world in order to make clear how little thought is given to the relationship between place and ecology in modern Western culture. It almost seems to be a blind spot in modern thinking. The question that arises is: At what point did the relevance of the place on earth in which we live disappear from the image we have of ourselves as moral beings? Has one of the effects of modernization been to alienate us from local sensitivities and responsibilities?

As modernity gathered pace, 'man' became an abstract notion, a bodiless character, defined with the aid of general concepts such as freedom and reason, free will and autonomy. In the course of the nineteenth century, this image of human beings became linked with an emancipatory understanding of history. With philosophers like Kant and Hegel as its most effective mouthpieces, the idea gained ground that the entire history of humankind until then should be understood as one long, upward development that had reached its greatest heights in Europe. What had occurred on earth before then was to be regarded as the prehistory of a rational humanity. Henceforth, Enlightenment philosophers believed, modern humans would establish their authority on earth by means of reason. This view of history provoked fierce opposition even at the time. Karl Marx claimed that the dominant vision was nothing other than that of the ruling class, hoping to camouflage its seizure of power. Nietzsche too made rather sarcastic remarks in response to this linking of morality and history, for instance in an aphorism entitled 'The "humaneness" of the future'. It begins, 'When I contemplate the present age with the eyes of some remote age, I can find nothing more remarkable in present-day humanity than its distinctive virtue and disease which goes by the name of "the historical sense"' (GS, IV, §337). The 'historical sense' to which Nietzsche refers (always between inverted commas) is the view of the human species that prevailed in the nineteenth century. His description of it as a 'virtue and disease' suggests that what the Enlightenment citizen regarded as worthy of merit was diagnosed by Nietzsche as a sickness.

In forming their ideas, Kant and Hegel draw on the heritage of classical antiquity. The linking of human dignity with freedom and rationality certainly goes back a long way. Stoic philosophers Epictetus, Seneca and Marcus Aurelius in particular argued convincingly that every individual human is uniquely valuable in a moral sense. They developed this principle into an ethics of human worth. But if I, with my terrasophical perspective, search for the attitudes of the Stoics to the kinds of activities that Nietzsche believes connect us with the earth – the production and preparation of food, for example, or gardening – and their cultural and philosophical expression, then it turns out that such earthbound activities were not included in ancient thinking about human dignity. They were assigned to slaves or to people from outside the *polis*. If we add to this the fact that in Plato's philosophy everything that has to do with the body and our earthly existence is placed in an unfavourable light, then it becomes clear how deeply anchored in Western philosophy and culture is the process Nietzsche calls the 'degeneration' of humanity.[31]

The ecological crisis forces us to think again about where the value and dignity of human beings lie. Contemporary Chinese philosopher and humanist Tu Weiming writes that we must not abandon the core values of what he calls 'the Enlightenment mentality' but instead situate them afresh in the context of the cosmos. Not human beings but the life of the community of earthlings as a whole, focused on a shared future, must become the coordinating framework. The values of life, freedom and the pursuit of happiness, values that sum up modern humanist discourse, are then given new content.[32] The Western anthropocentric world view, with humans as the centre of all assignment of value, gives way here to what Tu Weiming calls an 'anthropocosmic worldview', in which humankind is seen in a cosmic perspective but nonetheless allowed to retain its own unique value.[33] In Nietzschean terms a turnaround of this kind in world view can be described as a shift from a human-centred vision to one in which life itself, in its evolutionary development, is regarded as a starting point, looking out on an open horizon.[34]

Two questions for further research

In the twenty-first century the word 'local' has different connotations from those it had in the past as the village or small town, an area relatively isolated and closed off, with an economic and cultural identity of its own. Nowadays, all over the world, the local has become an arena in which the specific characteristics of local life are continually intermixed with influences from outside, forces that are attempting to absorb it into globalizing economic processes and planet-wide cultural and communicative trends. People are rising in opposition to this situation in more and more places. The local is deployed in support of new developments in the fields of economics, society, politics and culture. At one end of the spectrum, traditional local values and customs are the point of departure for a revival of politics of a nationalist, even fascist kind that still has deep roots. At the other, movements of social renewal are arising that make the local the central concern of an ecological vision of society and culture. In his resistance to patriotic nationalism, Nietzsche went in search of a radically new, non-modern connection with the earth, and in doing so he anticipated what is happening now at many places in the world. In the Global North and the Global South, more and more initiatives are striving for greater local autonomy, based on an ecological vision of the future. As Ezio Manzini puts it, a 'new sense of place'[35] is at the heart of multifaceted social renewal. Manzini has introduced the term 'cosmopolitan localism' for this current of thought, applying it to all movements in which local roots and backgrounds are connected with a striving for what Arturo Escobar calls 'a dynamic reinvention of the local and communal through a multiplicity of activities concerning food, the economy, crafts and care'.[36]

This brings me to my two questions for further research. First, how in the Western context can we reassess the value of the locally situated? After all, it is there that our primary connection with the earth lies. Secondly, how can the relationship between local life and the life of our planet as a whole be deepened and strengthened? How do they relate to each other and influence each other? I will look briefly at both these questions.

As we have seen, involvement with the local environment has been subordinated to the values of neoliberalism, to increasingly strict regulation of the open market and to the global exchangeability of goods and services. Daily life is thoroughly permeated with economic incentives. But the process of globalization involves more than the worldwide rise of neo-capitalist systems. The term also refers to a process of material and cultural exchange between places and people, close by and far away, an impalpable, polycentric affair that generates both hope of a better life and fear of the future.

A second structural factor concerns the changed influence of the state, modernity's greatest political invention. With globalization, the nation states have come under strain. At the top, their power is limited by international political bodies that in more and more places set the limits within which the state has to operate. At the bottom, a shift of power is underway towards regions and local government, with initiatives that attempt to breathe new life into local involvement. Here I look at those that concern ecology and lifestyle, examples being the tendency to promote locally produced food and to generate energy locally, the many forms of city farming and urban greening that are gaining ground, the rise of the Transition Towns movement,[37] the success of the Colibri movement[38] and so on. Citizens are feeling motivated to make their direct environment more pleasant to live in from an ecological point of view, and there is growing interest in sustainable development in the fields of housing, energy use, transport and nutrition, all reinforced by increasing knowledge of ecological issues. This involves exploring opportunities for greater economic independence and searching for a new balance between local and global interests.

I regard such initiatives as an expression of a new solidarity with and sense of responsibility for the particular place on earth in which we live. The combination of these two values, solidarity and responsibility, forms the core of what I call the indigenous mentality. I believe that alongside the aforementioned forms of nativeness, a typically Western variant could be developed. In the context of late-modern society, 'becoming indigenous' could mean working for the local environment on ecological grounds. Although not a native in the strict sense, by birth, you could become so by commitment, so that the place on earth where you live becomes your habitat, the area of which you are part (along with the rest of the community) and for which you take responsibility. The wisdom that is arising afresh in local communities can serve as a motor for a culture of local ecological innovation. In the context of high-modern Western society, 'becoming indigenous' would be a title of honour. This seems surprising, since the words 'native' and 'indigenous' still have the pejorative connotations they acquired in the colonial era. Some explanation is therefore required regarding the relationship between the present and the imperial past. But the term is above all at odds with the established values of globalization and neoliberalism, and provokes discussion of those values. Diverse local initiatives aimed at re-localization might grow to become a flexible network and at the same time a powerful movement that has a political force of its own and contributes to the bolstering of international cultural resistance to the negative effects of globalization.[39]

Regarding the second question, about the relationship between local life and the life of the earth as a whole, in our time we are aware that we share the earth with the rest of humankind. Even the most local ecological consciousness is pervaded by this planetary dimension. We might even say that ecological consciousness began when people started to take account of the condition that life on earth as a whole found itself in. The global character of the crisis began to work through at a local level, as people started to experience the consequences in their own environment. Ecological consciousness therefore automatically produced the insight that what we do is linked to the larger processes of life. Yet there is hardly any framework within which we can connect issues surrounding the strengthening of local life with the international politics of ecology. The world looks first to initiatives by international players, such as the climate conference in Paris in 2015, or builds on the strategy of the United Nations, which formulates measures that then have to be worked out in practice at a local or regional level. For the next decade or so the United Nations has put together a list of seventeen goals for sustainable development to give shape to its strategy worldwide.[40] Ought there not at the same time to be a powerful bottom-up movement that makes local communities, their interests and initiatives, an influential part of the process? The multifarious nature of local life and the diversity of local cultures could then be an important counterweight to the standardizing temptations of top-down strategies and create a fruitful dynamic between local and global initiatives.[41]

Becoming indigenous: A philosophical reflection

The expression 'indigenous people' has a history all its own. For centuries it was used to refer to uncivilized communities that needed Western civilization if they were to become human. Ethnology then removed the negative connotations from the term, turning it into scientific jargon and a field of research. The word 'indigenous' was at first used mainly for plants and animals. Later it became a way of indicating that a given group of people had held on to its connections with its pre-colonial culture and was now demanding an identity different from that of the powerful. The connection between culture and land was regarded as the most important feature of such demands. The English terms chosen to refer to these communities reflect this, beginning with 'natives', 'tribals' and 'Indians', since replaced by 'Native Americans', 'Aboriginals', 'First Nation peoples' and so on. Ethnologists concentrate mainly on the past, showing us what has been handed down.

I propose using 'indigenous' in such a way that it can also point to the future. 'Becoming indigenous' then refers to the creative effort to establish and preserve connections between local culture and ecological consciousness. In the context of late-modern society, becoming indigenous will then be an attitude to life and a strategy for experimental living. Whereas economic globalization is led by a logic of interests and profits, being indigenous nourishes a consciousness of local responsibility, from a planetary perspective. If we put ecological commitment first, the decisive criteria are no longer origin or blood ties but instead future-orientated responsibilities. The term 'local community' then stands for openness, as regards both the composition of its population and the nature of its responsibility. People who come from elsewhere can become just as indigenous to a specific place as people who were born and raised in it. There are advantages to bringing in people from other cultures, since they contribute knowledge and skills, customs and stories that may well enrich and expand existing ideas and habits. The debate about multicultural societies thereby gains a new dimension. The distinction between native and newcomer makes way for qualities such as commitment and responsibility. Where cultural plurality is deployed to aid local ecological awareness, it will prove of great value.

The second advantage of such a Nietzschean-inspired approach is that the goal of the indigenous will no longer be the restoration of the 'original' culture as a norm for the future. Genealogy is not aimed at restoring the past, rather it is a kind of therapy that helps us to let go of aspects of the past that hamper the development of an ecologically aware culture and to discover those forgotten or repressed values that we need to take with us when we focus on the earth in the future. The protection of indigenous customs and mores is not a goal in itself. We need to avoid romanticizing or idealizing the past, in whatever form, if we are to bring about a renaissance of this kind.[42]

6

Remain faithful to the earth
Towards an ecological cosmology

A changed perspective on the earth

The earth was really only properly discovered when astronauts showed what our planet looked like from space: a small globe surrounded by a thin biosphere that, as Dutch astronaut Andre Kuijpers put it, looked as if you could almost blow it away. Kuijpers was one of many astronauts who were deeply affected when they saw the earth rise before them from behind the moon. One of the first, James Lovell, who orbited the moon on board Apollo 8 in 1968, said that the earth looked like 'a grand oasis in the vastness of space'.[1] On seeing the photographs that same year, poet Archibald MacLeish spoke of a 'tiny raft in the enormous empty night',[2] while mythology scholar Joseph Campbell wrote in his book *Myths to Live By* of the earth appearing as 'an extraordinary kind of sacred grove, set apart for the rituals of life'. This vivid language indicates just how overwhelming, how awe-inspiring, how moving those first impressions were. The astronauts realized that the beautiful planet earth, from which they had come and to which they would soon return, was the only place in the immensity of the universe where they could live. 'Earth' came to mean our home: Homeland Earth. Cosmonaut Oleg Makarov, who took part in five missions of the Russian Soyuz programme, put it like this: 'Suddenly you get a feeling that you've never had before, that you're an inhabitant of Earth.'[3] Apollo 10 astronaut Tom Stafford noted, 'You don't look down at the earth as an American but as a human being.'[4]

Leaving the earth enabled the birth of a new consciousness. The earth was not just a special and exceptionally beautiful planet, it was home to all those who lived on its surface. The borders between countries, their national differences, seemed relatively unimportant. The crucial fact is that we live on planet earth as one species. This sense of connectedness is made all the more intense by the pitch-black cosmos against which the earth lights up. It prompts a feeling of cosmic loneliness. We experience a stark contrast between our bright, colourful home and dark eternity. Space travellers seemed to develop a personal relationship with the earth and all the life on it, recognizing its infinite value.[5] This has now been further reinforced by the fact that from space we can see the destructive effects of the ecological crisis and the depletion of the biosphere. During his second space mission, in January 2012, Kuijpers said that he could clearly see a contrast with his first. The bare patches in the Amazon region were larger, and the blue clouds of air pollution above several cities were thicker. He concluded that he was now truly looking at a problem.

The view from space makes us aware that the earth is both fragile and indispensable for humankind, all the more so because the information received from space travellers is supplemented daily by observations from a growing number of satellites. 'The effect of seeing the earth from outside is both salutary and terrifying,' writes Wilhelm Schmid in his book *Ökologische Lebenskunst* (2008). 'Salutary because it reduces to its earthly proportions the human world that we regard as the most important; terrifying because the earth turns out to be so lonely and brittle.'[6] The fragile beauty and life force of the earth have made us sensitive to the unique character of our planet. Going 'back to the earth' means discovering that the modernist image of it as a homogenous globe that can be captured in a single glance needs to be replaced by that of a patchwork of life forms and cultures. We need to exchange the neat globe that nineteenth- and twentieth-century scholars liked to place on their desks with the image of a multifarious, living earth.

The other discovery that has changed our perspective fundamentally relates to our understanding of time, to the way in which we locate ourselves on the axis of time and understand ourselves as temporal beings. This change was prompted by what we could reasonably call the greatest discovery of the past 150 years, namely that everything that lives on the earth is the result of an ongoing evolutionary process. Space travel changed our experience of space, while the theory of evolution, first developed by Charles Darwin – a product of the nineteenth century like Marx and Nietzsche – caused a profound change in our conception of time. Humans are evolutionary latecomers and as a species they belong to the mammal family. We need to understand the development of life on earth and of the human species as a complex event, a network of influences within which everything is connected. That connectedness is now under threat. It was climate scientists, geologists and biologists who sounded the alarm and made clear that we need to become more aware of being inhabitants of the earth. In this final chapter I first look at knowledge about the relationship between humans and the earth that has developed relatively recently. Then I place Nietzsche's ideas on the subject alongside these new insights.

The Anthropocene: The context of our time

It was not until the second half of the twentieth century that links were made between the current ecological crisis and human behaviour. Scientific discussions have since led to the introduction of two new scientific concepts, Gaia and the Anthropocene. Gaia is a name for the earth introduced by climatologist James Lovelock, who was eager to show that our planet should not be seen as one heavenly body among many but as a complex, living entity, consisting of a web of interconnected ecosystems, the earth as a whole being the largest ecosystem of all. With his analysis he demonstrated that the climatic balance on earth, a precondition for all life, is changing because of the worldwide increase in modern ways of living and will be seriously disturbed within the near future if people go on as they are. The title of one of his later books, *The Revenge of Gaia*, speaks for itself. As well as giving us information about how things stood on earth, Lovelock placed that knowledge in a genealogical–therapeutic perspective.

The other discovery is no less important. Under the leadership of Dutch geologist and Nobel Prize winner Paul Crutzen,[7] an international commission of geological researchers came to the conclusion that of all the forces on earth, humans are now the most influential. To mark that transition, they proposed adding a new era to the recognized geological epochs (Pleistocene, Holocene and so on). They called it the Anthropocene, an era in which the Anthropoi, humans, are profoundly altering the geological situation on earth. Crutzen and his team identified three sub-phases within the Anthropocene. The first runs from the Industrial Revolution, or roughly

1870, to about 1940. The second covers the period 1940–90 and is known as the phase of acceleration, characterized by explosive demographic growth, and by unprecedented economic and technological developments and their worldwide dissemination, a process we have come to call globalization. In this phase the ecological problems could be discerned, they write, but the holders of economic and political power paid them little heed, and they had little if any impact on public opinion. The third phase began in about 1990, when we started to become fully conscious of increasing human interference in 'system earth'. International summits, such as the UN climate conferences in Rio de Janeiro (1992 and 2012) and Copenhagen (2009), took up the theme. Along with 'Gaia', the concept of the Anthropocene has since been embraced by an interdisciplinary community of researchers from what are known as the hard sciences and from the social sciences. They define 'system earth' as a new object of study.[8]

These new concepts have been decisive in increasing our awareness that the relationship between humans and the earth has changed radically. It is indeed astonishing how the subject matter that Johan Rockström sums up in the phrase 'planetary boundaries'[9] has spread across the world. In no time at all, the conviction has grown that new problems are presenting themselves that deserve the attention of everyone who is involved in any human society and is at all sensitive to the demands of the future. The challenge these developments present has been taken up by philosophers too. From my perspective, philosophers who connect ecological problems with a critical analysis of the modernist concept of nature are especially relevant. Inspiring thinkers in this field include French ethnologist, sociologist and philosopher Bruno Latour, Belgian philosopher of science Isabelle Stengers and German philosopher of culture Peter Sloterdijk. Each of them contemplates the significance of recent developments for the current situation, and for our view of the world and of our own species. Latour writes,

> What is so ironic with this anthropocene argument is that it comes just when vanguard philosophers were speaking of our time as that of the 'posthuman'; and just at the time when other thinkers were proposing to call this same moment the 'end of history'. It seems that history as well as nature have more than one trick in their bag, since we are now witnessing the speeding up and scaling up of history not with a posthuman but rather with what should be called a post-natural twist![10]

Stengers characterizes the concept of Gaia as a radical 'breach' in modernity. Gaia brings about a thorough disruption not just of our living conditions but of our scientific thinking and the images we form of ourselves, of others and of the earth. What we refer to by the concept of 'Gaia' will not fit into our established framework of concepts and ideas, she claims.[11] The central values of autonomy and sovereignty on which our identity is based are disrupted by it too. We need to think about what it means to be confronted with a power that takes us beyond the boundaries of our concepts, as Stengers puts it, because Gaia presents herself as a transcendent entity of a secular nature, just when we thought we had conquered transcendency. Latour too calls Gaia a secular power of unprecedented greatness. The earth turns out to be an actor that we cannot comprehend as long as our ideas about actors and sovereignty are applied to people alone. Gaia is a new kind of actor, a sovereign power with which we have to learn to deal in all fields – scientific, cultural and religious, but above all political and economic.[12] Sloterdijk meanwhile calls the earth the only entity capable in the present age of exacting from us a categorical imperative that all people must adhere to if humankind is to survive. He refers explicitly to Nietzsche when he introduces the earth as the only moral authority that can still command the loyalty of the human species.[13]

As I have said, I interpret the thinking of these philosophers as contributions to a genealogical analysis of the relationship between humans and the earth in our time. In their reconstruction of that relationship, all three focus on the threat represented by the current situation and on the symptoms by which that threat manifests itself.

Latour and Stengers point first of all to the feelings of panic and impotence that this new information brings about in large sectors of society, and the sense of being disconnected. There is a huge gap between the facts that science gives us about the state of the earth and our capacity to respond to them. Our feelings of ignorance and powerlessness seek an outlet and encourage the unquestioning acceptance of bleak scenarios concerning the future of humans on earth. Where the threat is experienced so powerfully that it exceeds human comprehension, people become subject to fantasies about the end of the world, and easy prey to messengers who try to channel these feelings into new depictions of the future that are mythical and apocalyptic. How should we react to this? How can we prepare for it? How can we find a way of looking at the future such that people will be able to identify with the existing state of affairs without losing heart? Latour is convinced that the hard sciences alone cannot fulfil that task. He writes that we need to develop other ways of relating to the earth that can give direction to our lives. Here he sees a new task for the social sciences, politics and art, and for philosophy.

That the scientific information has caught us by surprise and thrown us into confusion makes something else clear as well. The authors explain the resulting 'cold panic' (as Stengers describes it) and the sense that we lack, cognitively and emotionally, the means to respond effectively as symptoms of the fact that modern culture has done nothing to prepare us for this situation. The way in which the dividing line between nature and culture has developed in modern thinking is revealing itself as an obstacle to understanding. We still draw a radical distinction between the natural order on one side, which the hard sciences investigate, and on the other the order of human affairs, of freedom and autonomy and the societies we have constructed based on them. This distinction has penetrated all levels of our culture. It obstructs our view of what is really happening between humans and the earth. The new insights therefore reveal not only that, as Nietzsche put it, one of the earth's diseases 'is called: "Human being"', but that we cannot overcome those diseases until we provide ourselves with the right intellectual frameworks. An ecological criticism of science will therefore need to become a fundamental part of our criticism of modernity.

Finally, Latour and Stengers claim, information about the deteriorating state of the earth feeds a sense of urgency. Recent analyses are littered with crisis terms: climate crisis, fish stocks crisis, rainforest crisis, biodiversity crisis and so on. The entire list quoted by Rockström, all the rhetoric that is continually being supplemented by fresh scientific evidence, is a symptom of the seriousness of the current situation that calls for a transformation, for a 'cure' as Nietzsche puts it.

In the genealogical diagnosis of modernity, the relationship between humans and the earth appears as a trial of strength of unprecedented magnitude, a contest that may be decided against us if we do not change. The current situation confronts us with a radical choice that Latour expresses as 'Modernize or Ecologize!' Anyone who understands this diagnosis of our time will break out of the boxes inside which modernity has placed nature and culture and take steps towards an ecological approach to humankind and the world.

Against the background of these analyses, Nietzsche's call to remain faithful to the earth seems intended for our time. With the help of the aforementioned writers I want to investigate further Nietzsche's exploration of the earth and suggest a number of approaches that I believe can lead to a fruitful exchange between his ideas and contemporary thought. I focus in my reflections on the two poles that are central to both: the earth and the human species. What do we learn if we put the Gaia hypothesis and Nietzsche's vision of the earth side by side? In what ways do Gaia and Nietzsche's earth differ, and in what ways do they complement or challenge each other? I will then address Anthropos, the figure of the modern human who plays a central role in the Anthropocene, and bring him into conversation with Nietzsche's figure of the overman. Can the overman and Anthropos complement each other, or are they mutually exclusive?

Nietzsche's earth

Let us return to the way in which Nietzsche presents the earth in *Thus Spoke Zarathustra*. It is imagined as the source of life with which humans must reconnect. But the earth has two faces. One is the natural world around us, with which we are connected body and soul. At the same time 'earth' refers to a reality that is not yet here and that, Zarathustra suggests, will reveal new aspects of itself with every step we take in its direction. The earth is 'open', beckoning to us from the future. Zarathustra says that the new experiences and insights we acquire in our quest for the earth will be broken open again and again to reveal different prospects, not only during the lives of people now on earth but generation after generation. The earth therefore creates a bond, a relationship in which people discover step by step what it means to live as a human being. Because Nietzsche's earth becomes a figure, a name, a sign, getting people moving and at the same time leaving the horizon open and continuing to advance, he eventually describes being human as setting out on an adventure, an experiment that is never completed and never reaches its goal. In Nietzsche's view of the earth, becoming human is a continual, ongoing process, which mobilizes and draws upon all human capacities and desires but never makes them immutable. Setting out towards the earth involves our entire existence.

Gaia

Under the name Gaia, the earth sciences introduce new climatological, geographical and biological knowledge about our planet that helps us to be far more precise about the nature of the earth as a living entity. The Gaia dossier, so to speak, is supplemented every day with new facts about the current state of life on earth. This knowledge has made the disastrous impact of modern humans visible, and it paints a gloomy and threatening picture of the future. Lynn Margulis[14] and later James Lovelock have made clear, however, that humans will not destroy life as such. Life on earth has developed over millions of years and is far tougher and more extensive than anything we could destroy. Our modern way of life damages the earth as a habitat for humankind and thus threatens the continued existence of our species, but life in a larger sense will continue. Margulis and Lovelock have been drawn into fierce battles against the arrogance of those who thought they had life on earth in their hands, and in that respect Latour and Stengers agree with them. Latour, for example, takes the idea of the earth as a habitat and applies it to culture. In that light he looks at how various cultures have imagined the relationship between humans and the earth. The modern figure of Gaia should not be confused with the nourishing and caring mother honoured by indigenous cultures, he says. Today's Gaia no longer has the face of a caring mother. She now appears as a furious goddess, hitting back to show people that she will not tolerate their derision. If we want to capture our relationship with the earth in images of Gaia, we should choose not the mother–child relationship but that between two adult actors, a relationship that has become strained because people have ignored the earth-in-action and reduced its status to that of a dead object that can be plundered. Gaia's reaction makes clear that she is alive and that we must take her seriously in a new way. The earth is not an object but a partner. Like anthropologist Philippe Descola, Latour extends our knowledge of Gaia into the cultural field and wonders which new cosmology, which cultural, ethical and political images and representations, might persuade people to change their attitudes.

The current alarming situation, characterized by a long list of crises, will not change of its own accord. In a wide variety of ways, the same conclusion is reached time and again: the future looks bad, and in fact we may already be too late to change it. We need to learn to live with the prospect that many generations to come will be confronted with an inheritance that does not allow them a positive outlook. Such diagnoses have led Latour and Stengers to believe that the first task facing the social sciences and philosophy is to confront people with this reality, to shake them awake, to make clear to them its seriousness and urgency, and to stimulate ways of imagining and depicting that are independent of science, so that people can connect with the situation not just cognitively but emotionally and spiritually. Scientific information alone will not reach the majority of people, let alone prompt them to act. If we really do have to change, then other areas of social life – art, media, film, theatre, religious and ethical spheres of influence – will need to be involved. In short, what is presented to us under the name Gaia introduces a vision of the earth in which the planet distinguishes itself in the universe as a living entity of which the life of our own species is only a small part, and one that has entered a historical phase in which it becomes clear that severe disturbances to human life will affect many generations after our own.

Gaia and Nietzsche's earth: Three differences

If we now lay the two visions side by side, three differences stand out. First, the Gaia concept offers us a macro-perspective on the relationship between humankind and the earth, whereas Nietzsche starts at the other end, with the question of how the individual and the local community experience their place on this planet.

A second difference concerns science. Gaia is the product of specific scientific knowledge about the earth. This becomes the frame of reference for all thinking about humans and the planet. The most striking effect, as we have seen, is that the future becomes bound up with notions of crisis, threat, downfall and defeat. Nietzsche had a different attitude. As noted earlier, he unquestionably took on board insights from biology and evolutionary theory, but in his view any science that approaches the earth as an object is of its nature one-sided and can never make room for all the facets of humanity brought to the fore by contact with the earth. The earth shows itself as powerful, magnificent and capricious, as an explosion of life forces that arouse all kinds of feelings in us, including joy and generosity, for which the sciences have no time. Science is important, Nietzsche believes, but it is not the only source of knowledge about the earth. He therefore formulates the need to absorb it into a wider view, with room for other insights and experiences in our contact with the earth. The Nietzschean perspective gives the individual and the community a way of thinking that can work as a counterweight to our all too straightforward applications of science to everyday life. Against the darker background of Gaia, Nietzsche's earth can nourish our desire to tap the positive healing and vitalizing energies available to us as a result of contact with nature. In a Nietzschean approach, science is not ignored, but space is created where the paralysing effects it sometimes produces can be addressed and absorbed into a broader perspective. The comparison between Gaia and Nietzsche's earth makes relevant once more, in a new way, questions about how scientific knowledge and existential knowledge can be interlinked and how they relate to each other. Throughout his writing, Nietzsche advocates the development of a 'postmodern' cosmology, which I call terrasophy, as a framework for a new philosophy of existence in which humans are no longer defined exclusively on their own terms. If we take Nietzsche's thinking further, it supports my assertion that cosmology ought again to become a core discipline of philosophy. In almost all cultures – Greek, Roman, Oriental, indigenous – cosmology offered a framework within which people could broaden their image of themselves.[15]

Thirdly, these two characteristics of Nietzschean thinking – starting with the individual in a local context and refusing to take the scientific way of thinking as the ultimate frame of reference – have their roots in an idea that I see as the basis of Nietzsche's philosophy of the earth, namely that humans and the earth must be defined in relational terms and can reveal their meaning only in an active process of relating to each other. He saw the earth as both the origin and the purpose of human existence, its alpha and omega, to use religious terminology. Being human means returning time and again to that origin and then, enriched by the experiences and insights gained there, tracing out a path to the future. To become human is to set out on a road of which the earth is the direction and goal. The earth is an empty signifier; in other words, our existential knowledge is the fruit of interaction. Instead of taking god as alpha and omega and explaining human existence in terms of an existential relationship with the divine, as Kierkegaard did, for instance, in that same nineteenth century, Nietzsche advocates discovering what changes when we take not god or reason or humankind but the earth as the framework from which we derive the meaning and significance of our existence. Nietzsche put this into practice, for example by setting aside the images of nature he had acquired and going out into the natural world, walking for hours and opening himself up to all the impressions that came to him there. It is an attitude I would like to call 'primary naturalism'.[16]

Nietzsche sought observations that were as unmediated as possible – in that sense you might call him a phenomenologist *avant la lettre* – but he liked to focus his attention on those manifestations of nature that have become emotionally charged as a result of morality and religion, and banished from human experience. Think of his fascination for the predator, for things that strike us as violent and uncaring in life in general. Think of his brutal image of nature, of what he mockingly called his 'black Rousseauism', to distinguish it from the image of adorable natural life presented and exalted to a norm by Rousseau and before him Epicurus. Nietzsche's earth-as-origin also has the characteristics of a wilderness that are so admired by deep ecology and nature conservation, a natural environment that excludes humans. We have seen how Zarathustra seeks contact with the animals and plants, the mountains, the sun and the sea, and how he approaches humans as creatures that have yet to learn what earthly life means.

The overman

As with Nietzsche's earth, in relation to the figure of the human we find in Nietzsche a narrative tension that sets out both the present and the future. But whereas in the case of the earth one and the same term refers both to the present day and to our distant prospects, when he writes about humankind, Nietzsche gives the future the name 'overman'. What, then, is the figure that represents humankind as it is today? When he is referring to modern humans, the representatives of modernity made flesh, as in *Thus Spoke Zarathustra*, he often writes simply of 'the human being'. As we saw in Chapter 3, Nietzsche presents on the one hand 'the last human being', whom he calls 'the most contemptible human', and on the other 'the higher men', who are visibly searching. He portrays taking 'steps to the overman' as freeing oneself from the identity of the modern human and taking leave of the city. But alongside this figure he constructs yet another image, the most important component of which is the body, asymptotically presented as a formless, chaotic, complex interplay of urges and instincts, of emotions and fantasies, lusts and 'wills' that need to be imagined as completely naturalized, which is to say stripped of all the human judgements that are attached to them. This is the body as radically open potential, without prescribed orientations. Nietzsche rids human bodily potentials of moral judgements in particular, because it is they that obstruct him the most in his effort to gain access to all the powers humans are capable of having, including the most violent, and to interpret them. I see the sometimes shocking passages in his work about human instincts and passions as hermeneutic attempts to visualize this brutal stratum of human existence, in defiance of the pressure of prevailing morality, and to make it philosophically accessible. By using the word 'brutal' I do not mean to suggest that Nietzsche wants to glamourize the brutality of violence – he does not – and I am aware of the risks that the use of such a word brings with it. But I have chosen it for the sake of its connotations: powerful, raw and indeed ill-mannered and contravening propriety.

Nietzsche puts forward the idea that this brutal, impulsive potential contained in the body – one of the meanings of 'the not yet determined animal' – might have any number of different outcomes.[17] The entire spectrum shown to us by animals, from predator to obedient sheep, from chimpanzee to bonobo, is therefore in theory open to us.[18] Which is why Nietzsche immediately connects this to the fact that, unlike other organic beings, humans always need moulding and stylizing. Becoming human means giving form and style to bodily potentials. The core question then becomes: What form and what style? If we now take this figure, the 'brutal' body, as a correlate of the overman, then the two poles of the human narrative tension become visible. Nietzsche challenges us to undergo a dual transformation: dare to free yourself from the straitjacket of the moral, religious, bourgeois, proletarian forms of modern humanity in which you were socialized and with which, more or less wholeheartedly, you have identified. Seek to restore contact with a naturalized body. At the same time, looking outwards, reform, transform and style your freed potentials in the direction of the earth. In *Thus Spoke Zarathustra* these transformative moves are called 'the steps to the overman'. For humans the overman is 'the meaning of their being', and 'the overman is the meaning of the earth'.

The body also provides Nietzsche with an evolutionary dimension. We have an overly short-sighted and narcissistic image of ourselves if we leave out how in the body both inorganic and organic evolution work through in us.

Let us now compare this human who, to adapt Nietzsche, dances on a rope fastened between the brutal body and the overman,[19] with Anthropos. To whom does the Anthropos of the Anthropocene refer?

The Anthropos in the Anthropocene

In the discourse of the Anthropocene, Anthropos is the name for the generations of humans who, over more than a century, have developed and disseminated a certain way of living on earth. With the introduction of the term 'Anthropocene' to mark a new phase in evolution, the expression 'humankind' has gained a material dimension of meaning. Having once been an abstract, philosophical idea, the bearer of a universal morality, humankind is now becoming first and foremost a historically demonstrable collectivity. From now on, all human beings are bound together by a historical fate and destination. At the same time, the abstract notion of humankind is increasingly being called into question. Does it still make sense to use the word 'humankind'? The notion of the ecological footprint, now broadly accepted, makes empirically visible the fact that different lives diverge markedly in the ecological burdens they represent. In this sense there is an immense difference between the American way of life and that of Europe, or of India, or sub-Saharan Africa, and within those populations there are big differences again between city dwellers and those who live in rural areas. Other studies make clear that social differences ought to be included in the analysis too. Poorer population groups not only have less of a share in modernization and less of a say in that process, worldwide they are affected first and most severely by the deteriorating situation. They bear most of the costs of the destructive lifestyles enjoyed by the better off. Another approach would be to focus on those mainly responsible in the field of economics and to name them publicly.[20] So in various ways the category 'humankind' acquires different faces and we start to see the interests that different groups have in maintaining or changing the status quo. When it comes to the questions raised by Gaia, we are a long way from a situation in which our species can be proposed as a single subject with a shared, universally recognized morality. From a historical and sociological point of view, 'humankind' as a whole is for the time being a fiction.

The geologists who invented the term 'Anthropocene' did however come up with suggestions as to what the human species needed to do. They called upon it to form a kind of world government that would take responsibility for decisions and for turning the tide. They took the United Nations as a model, imagining it transformed into an effective governing body. They are not alone in having such ideas, but their vision is controversial and has provoked fierce reactions from the opposition camp, where many say that such a top-down model shows nothing has been learnt from history and warn of the dangers of enforced unity. Furthermore, it is an approach that risks giving the powerful even more power and serving the interests of the strong. Precisely because such major interests are at stake, it could lead to a new form of dictatorship, a global dictatorship this time.[21] The Anthropocene creates a situation more like the Tower of Babel, says Latour, than a global conversation guided by shared values and universal assumptions. He adds that we may be on the eve of a new war, because interest groups will take tough positions in the continual polemic and struggle that will arise, and they will not recognize a neutral umpire. The interpretation of universal moral principles will become a matter of dispute too. In order to steer things in the right direction, Latour believes, we need to invent a new diplomacy that continually invites all the relevant parties to the table, including Gaia, and works step by step to build a peaceful future. He calls this strategy 'compositionism'.[22]

The overman and Anthropos: An important difference

There is a crucial difference between the notion of Anthropos and the Nietzschean perspective. Nietzsche never addresses his appeal to humankind as a collective. The approach he advocates begins at the grass roots, with direct experience and the immediate environment. The power of the transformation arises from an event that is both internal and social. On these foundations of personal experience and local involvement, we must build morally and politically. We have seen how sceptical Zarathustra is about the leaders of church and state who want to take people to all kinds of places but not to the earth. Zarathustra speaks to political leaders too, primarily about their own behaviour and motives. From a Nietzschean perspective, existential commitment – a personal involvement with the earth – is the basis and precondition for all real change. The starting point lies in a change in mentality that makes people aware of the current situation and causes them to look at themselves, at their behaviours and desires, their beliefs and expectations with fresh eyes. As we have seen, Nietzsche connects this change at a personal level with a cultural shift, which is then extended into the domain of education and childrearing. Personal, local and cultural change will drive social change. Beyond the boundaries of political power blocs and across lines drawn by nation states, Nietzsche's overman perspective calls for new forms of kinship and solidarity. Commitment to the earth leads to a new form of planetary solidarity. In Latour's vision the contemporary era confronts each of us with a choice: for or against the earth. He writes that two new collectives will arise, that of what he calls Earthlings or Earthbounds, who have connected themselves with the earth, and against them a collective made up of beings he calls Humans. In the spirit of Nietzsche, therefore, he reserves the latter term for creatures who hold on to their modern identity and way of behaving, refusing to acknowledge that the relationship between Anthropos and the earth has changed radically.

On the way to 'the great health'

I believe the shortest way to sum up what Nietzsche presents us with in the person of Zarathustra is to call him an inhabitant of the earth. Nietzsche gives the overman the title 'meaning of the earth' and a task: 'remain faithful to the earth'. His entire philosophical project can be seen as a personal yet radical attempt to give those simple words the explosive force that they gradually acquire for him. Might we be able to refer to the people who in our own time take 'steps to the overman' as indigenous people of a new kind who, where they live, always have the earth in mind? Are they 'cosmopolitan natives', so to speak? After all, in the final analysis our time calls for all people, right where they live, to become aware of their planetary responsibility, and what this means will need to be spelt out time and again. The way to this future will be long, and they will meet with considerable resistance, but those who choose to set out on this path will find inspiration in Nietzsche, who in his own quest, aside from moments of despair, disappointment and discouragement, expresses the enthusiasm and joy that the journey towards 'the great health' has brought him.

> *The great health* – Being new, nameless, hard to understand, we premature births of an as yet unproven future need for a new goal also a new means – namely, a new health, stronger, more seasoned, tougher, more audacious, and gayer than any previous health. . . . And now, after we have long been on our way in this manner, we argonauts[23] of the ideal, with more daring perhaps than is prudent, and have suffered shipwreck and damage often enough, but are, to repeat it, healthier than one likes to permit us, dangerously healthy, ever again healthy – it will seem to us as if, as a reward, we now confronted an as yet undiscovered country whose boundaries nobody has surveyed yet, something beyond all the lands and nooks of the ideal so far, a world so overrich in what is beautiful, strange, questionable, terrible, and divine that our curiosity as well as our craving to possess it has got beside itself – alas, now nothing will sate us any more! After such vistas and with such a burning hunger in our conscience and science, how could we still be satisfied with our *present-day man*? (GS, V, §382, italics in original)

Afterword

How to continue in Nietzsche's footsteps

In Part II of this book I have attempted to create a dialogue between Nietzsche's vision of the earth and the view of our planet as seen by astronauts, along with a number of insights from contemporary science and above all from Western philosophy. What they have in common is the conviction that drastic changes are announcing themselves in the process we call life on earth, with major consequences for human beings. Nietzsche did not have access to contemporary science of course, but he did have a sharp and critical understanding of the foundations of modern culture, meaning the culture that continues to impact profoundly on all life on our planet. His philosophy can therefore prompt us to become more acutely aware of the issues we face and help us to set out a way to the future. Nietzsche first mentions Zarathustra in a note headed 'Noon and Eternity. Signposts to a New Life'.[1] This provisional title helps to clarify his intentions in writing *Thus Spoke Zarathustra*. To end, let me offer a few signposts of my own that I have derived from my meeting with Nietzsche.

How to look at the future

I regard it as extremely important, for our own age especially, that Nietzsche sketches prospects for the future based on positive values, and that he ranks 'great health' as the highest of these (rather than freedom, or power, or nihilism). In the present context I take great health to mean a situation in which humans and nature reinforce each other in their vitality and resilience. We may be no less critical of our own time than Nietzsche was of his, and no less gloomy about the way society is currently set up, but such responses are put into perspective by the realization that, instead of seeing ourselves as the last generation, we would do better to understand ourselves as transitional figures on the way to a new future. Many will of course ask: Is there in reality a new future for us? With the coming of the Anthropocene, the conviction has rapidly taken hold worldwide that we are already too late, that even a politics of radical change will lack the capacity to turn the tide, that catastrophe is unavoidable and the only proper response is to regard ourselves as terminal patients preparing for extinction, who need to learn to draw upon the resources available to individuals facing imminent death.

This is the attitude adopted by Pablo Servigne, Raphael Stevens and Gauthier Chapelle, for instance, in a new science they call collapsology.[2] Although few would go as far as they do, to the question of whether humankind will ultimately survive we increasingly hear negative or sceptical answers, especially from young people. The future that is advancing upon us fills us with dread. I believe that Nietzsche's great strength is that he refuses to position himself as the owner of the future, in fact it is precisely on that point that he resists Hegel's idea of the 'end of history'. Hegel claimed an ability to describe that end, but surely today's forecasters are making exactly the same claim, only with the opposite outcome: extinction rather than the victory of the rational human. The current situation is of course very different. The calculations and predictions of the science on which our gloomy view of the future is based are of a quite different kind from Hegel's speculation – and we cannot take them seriously enough – but can we pin down the future on that basis and claim ownership of it? Nietzsche is passionately opposed to any such attitude.

For Nietzsche the future is a direction in which we need to move in the present, a longing for people who live differently and precisely for that reason represent a source of hope. Our image of the future will change depending on how radically people engage in this process of transformation. The strength of Nietzsche's position derives from the fact that on the one hand he is as radical as anyone in standing by the guiding principles that we must be who we are, live our own lives, be led by whatever truly does us good and makes us happy. He avoids any temptation to moralize. On the other hand he leads us to the recognition that this highly personal orientation can reach its goal only if people start to behave as what they are: links in the larger process of life, which has not by any means been explored in all its most hidden, sometimes mysterious, potentials. What affects me most is that in his striving for freedom and for a principled lifestyle, Nietzsche takes as his starting point the restoration of his bond with nature. He makes of this a personal experiment, from which comes his commitment to a different future from that to which modernism would lead. He becomes increasingly sensitive to the happiness of animals, to the circular rhythms of nature, to the beauty of the earth and the insight that we humans are interwoven with life on earth as a whole, in countless ways.

Making himself a participant in life on earth in the largest sense was therefore not an abstract theme for Nietzsche but a guiding principle of daily application. As we have seen, there arose in him an intense desire for the 'man of the future' whose coming 'makes the will free again, which gives earth its purpose and man his hope again' (GM, §24). Can we keep that hope alive and nourish it without lapsing into naive optimism or passive quietism? The best way to do so, Nietzsche seems to suggest, is not to be led by fear and pessimism, even if sometimes we can see no way forward and are close to being overwhelmed by dire reports about the future. We need to have faith that coming generations will discover opportunities and set out along roads that today's generations – myself included – cannot see. 'At this point just one thing is proper,' Nietzsche writes. 'Silence: otherwise I shall be misappropriating something that belongs to another, younger man, one "with more future", one stronger than me' (GM, §25). Giving one's all in the present, unconditionally, with faith in the future, is the first of the signposts I have been shown by my encounters with Nietzsche.

Should we make of the earth a city or a garden?

Nietzsche regards the earth as a living planet that will change continually as a result of human activity. He writes with approval that humans are designers by nature, and pacesetters of change. As I have said, Nietzsche is a process thinker in this respect, who characterizes human intervention in life on earth as a series of experiments that need to be carried out more deliberately in future. I regard the anti-technology inspiration that has been attributed to Nietzsche in the wake of Heidegger as debatable at the very least. It is true that we know better than ever how experimentally humans have set to work, and see more clearly than ever the destructive effects of those social and technological experiments. But in contrast to the technophobe prophets it seems entirely in the spirit of Nietzsche to say that the outcome of human experimentation remains unpredictable and will be dependent on what people decide to do. It is precisely this that makes the questions at the heart of his ecological philosophy so important. By which values do people allow themselves to be led in their experimentation? Do they realize that only the changes that work to the good of life as a whole are good for human beings, and that every essential transformation they bring about will have to be accompanied by an inner metamorphosis and a more principled lifestyle?

One of the most recent large-scale, apparently directionless experiments that humankind is currently undertaking is the move to the cities and the concentration of huge numbers in vast metropolises. Predictions suggest that by about 2050, 75 per cent of people will be living in urban environments.[3] This is a development with enormous consequences. It drives the move towards large-scale agriculture, the need to transport food over long distances, the increasing uniformity of chains of consumption and investment in the technologies that make all this possible. Cities have become places where residents are cut off from the natural environment and have often lost contact with nature altogether. It is important to realize that if this continues, by the end of the twenty-first century the dominant human type may well be a resident of one of those megacities, born and bred in what is at this point a highly polluted environment. Furthermore, big cities are becoming increasingly similar, all featuring the same high-rise buildings, the same retail chains and the same busy arterial roads around residential areas. We are starting to realize how much damage all this is doing to the earth.

Nietzsche's call to make the earth more like a garden raises critical questions about the way urbanization is proceeding. Cities too must fit within the larger whole of the earth's ecosystems. More and more citizens are realizing this. There is no getting around the need to place the relationship between the city and its rural and natural surroundings high on the agenda in this century, and to start to think from the perspective of the earth once more, even if this involves very different problems in the new cities of China and India, or in Africa, from those we will encounter in America or Europe. This is the second signpost, and it brings us immediately to the third.

A new sense of place: Locality, proximity and plurality

By adopting the eagle's cosmic view, Nietzsche discovered how varied and multifaceted life on earth appears. Multiplicity and diversity emerged as core values in life, whether in nature or in the realm of culture. Organic life exemplifies them, as does the diversity of human cultures. In the context of globalization especially, it is important to describe explicitly the vital part played by diversity, also in a geological sense. This has led me to ask for attention to be paid to the place on earth where each of us lives, and to appeal for a new interpretation to be given to local connections and responsibilities. We have seen how important the everyday experiences of eating, gardening and above all being outdoors became for Nietzsche, along with the local customs and mores within which they were embedded. He sought renewal and pursued his desire to find out what living in contact with the earth might involve. I recognize a similar longing in thinkers who advocate the promotion of local, relatively independent communities and an approach in which all the vital necessities of life – food, housing, transport, energy, clean air, a relationship with the surroundings and so on – are interrelated and, in combination, made more ecologically sustainable. The values 'organic', 'local' and 'building new communities' go hand in hand in this vision, and are inextricably linked.

Ezio Mancini makes a 'new sense of place' the core of a multifarious movement of social renewal. What he describes as 'cosmopolitan localism', saying it is 'capable of generating a new sense of place', is intended to describe this approach.[4] It is an expression that covers all movements in which local roots and profound thinking are combined with efforts to bring about, in the words of Arturo Escobar, 'a dynamic reinvention of the local and communal through a multiplicity of activities concerning food, the economy, crafts and care'.[5] A similar vision can be found in the work of sociologist Arjun Appadurai (*The Future as Cultural Fact*, 2013) and ethicist Richard Evanoff (*Bioregionalism and Global Ethics*, 2010). When they use the term 'local' they are not referring to its traditional associations with the village, the small town or an area that is relatively isolated and closed off from others, with its own economic and cultural identity. They stress that these days, all over the world, the local has become an arena where the characteristics of life at a specific locality are continually being combined with influences from outside that attempt to absorb that life into globalizing economic processes and global cultural and communicative trends.

Manzini and Escobar resist this dynamic and justify doing so based on an ecological vision of the future. 'Cosmopolitan localism produces a new model of well-being: a well-being in which a major role is played by the recognition of how much socio-cultural and environmental contexts can contribute to people's quality of life and to the resilience of the overall society,' Manzini writes.[6] Might cosmopolitan localism, which takes planet earth as its frame of reference, become a pioneering movement in this 'self-doubting present day' (GM, §24)? Might it help us to flesh out Nietzsche's idea of turning the earth into a garden? It is along those lines that I intend to pursue my thinking further.

The way to a radical humanism?

The last and perhaps most important signpost for me concerns whether Nietzsche's philosophy can be interpreted as a search for a new humanism. I see many points of departure here. After all, the search began with the posing of acute and critical questions of humanism, which came into being along with modernity and stimulated the rise of a culture in which the West designated itself master of the universe and proprietor of the earth. In Nietzsche that criticism never led to hatred or contempt for humanity, or to the abandonment of human aspirations to make the world better. He introduces no actor to replace human beings, no god or other superhuman power that might save us. In this sense the figure of the overman falls within the tradition of humanist attitudes to the world and to history, in fact it radicalizes humanist principles. But the task that Nietzsche gives the new humanism is to adopt aspirations that go beyond the happiness of any one individual and elevate life on earth in its full extent to a vista and a value. This task acquires a new significance in the Anthropocene era, in which humans are discovering that they have been shaping the world for some time, without being aware of the fact. In our hands, the earth has changed. It is not the stable planet it once was. The Holocene, the epoch in which the effects of human activity on earth were negligible, is indisputably over.

The earth, shaped by modern humans, is hitting back. Disturbances in nature demonstrate this daily. We are discovering that our hold on the earth is far greater than we imagined, yet at the same time both control and consciousness are lacking. Behind the existing order of things lurks chaos, not in the metaphorical sense that Nietzsche gave to the word but in concrete ways of which we are all too aware. In my preface to this book I indicated that intense contrast experiences are the result, in which people become so deeply affected that they can no longer continue living in the normal manner. Contrast experiences can paralyse us, but they can also create the space to go in search of new ideas and new values that may point the way ahead. They connect personal questions with the call for a profound cultural shift, for a revaluation of all values. Nietzsche's figure of the overman brings together both aspects of that shift: the move beyond human-centred thinking and the creation of a future in which real individuals can develop into responsible inhabitants of our planet, sharing it with all other forms of life on earth.

Notes

Preface

1 For the suggestion that our experience of the ecological crisis can be interpreted with the help of work on the contrast experience I am indebted to Renske van Lierop and Christa Anbeek. See bibliography.
2 BGE = *Beyond Good and Evil*. A full list of abbreviated titles can be found at the start of this book.
3 Gary Shapiro, *Nietzsche's Earth. Great Events, Great Politics*. (Chicago & London: University of Chicago Press, 2016). I turn my attention to it in an 'Intermezzo', to be found between Part I and Part II of this book. See also Adrian Del Caro, *Grounding the Nietzsche Rhetoric of Earth* (Berlin & New York: De Gruyter, 2004).
4 An unmistakable echo of Nietzsche's cultural criticism and praise for the earth can be found in literary works including André Gide's *Les nourritures terrestres* (Paris: Gallimard, 1897) and in writing by French philosophers Luce Irigaray, Gilles Deleuze and Felix Guattari, to limit myself to the French reception of Nietzsche. Heidegger likewise developed his own interpretation of Nietzsche's philosophy of the earth, and it has since been extremely influential.
5 Faber, Malte and Reiner Manstetten, *Philosophical Basics of Ecology and Economy*, Trans. Dale Adams, London: Routledge, 2009, p. 149.

Chapter 1

1 Letter to Gustav Krug dated 4 August 1869. Translation taken from *Selected Letters of Friedrich Nietzsche*, 2nd edn, ed. and trans. Christopher Middleton (Cambridge, MA: Hackett, 1996).
2 See also https://ndpr.nd.edu/news/23816-a-nietzschean-bestiary-becoming-animal-beyond-docile-and-brutal/
3 English translation taken from *Selected Letters of Friedrich Nietzsche*, pp. 163–4.
4 English translation taken from David Farrell Krell and Donald L. Bates, *The Good European. Nietzsche's Work Sites in Word and Image* (Chicago & London: University of Chicago Press, 1997), p. 122.

5 English translation taken from http://nietzschehaus.ch/exhibitions-and-events/exhibtion/
6 Opinions differ regarding how best to translate Nietzsche's term '*Übermensch*'. 'Overman', used by Adrian Del Caro in his translation of *Thus Spoke Zarathustra*, captures the idea that being human is a continual process of reinterpretation and of transcending oneself. No English translation has been found that satisfies everyone. See also note 28 in Chapter 3.
7 Rudiger Safranski, *Nietzsche: A Philosophical Biography* (London: Granta, 2002), p. 221.

Chapter 2

1 The notes Nietzsche made in this period are often schematic and cryptic, or formulated in a very short, truncated fashion. There are, for example, some passages in which he suggests a direct link between what a philosopher eats and what he then goes on to assert. Those passages are sometimes interpreted as early portents of Nietzsche's approaching madness. Colli finds this infuriating, since he believes they in fact signal Nietzsche's desire for a new way of looking and the pursuit of new intuitions (1996, pp. 151–2). Often Nietzsche's notes seem no more than inconsequential scribbles, not intended for publication. Yet that is exactly what makes them so interesting, Colli says, because it is precisely the difference in style that introduces a new train of thought. Hence they are different from passages Nietzsche intended for what he calls 'my readers', which have been further refined and given a stimulating, challenging and artistically stylized form. We have already seen a number of examples. Many of his avowedly experimental texts, by contrast, seem to have been written only for himself. They are private. They are notes that might be picked up again later. One note he made in this period reads, 'I don't respect readers any longer: how could I write for readers. . . . But I note myself down, for myself' (NF-1887, 9[188]). This is one of the notes that gave Colli the idea that the distinction between what was intended for publication and what was not might be important for the interpretation of Nietzsche's ideas. He even goes so far as to suggest that Nietzsche's later work as a whole should be read as the combination of two different sorts of text, one written with an eye to publication and shaped accordingly – including many 'try-outs' of passages that were not yet publishable, since Nietzsche never ceased to experiment with style and vocabulary – and the other sort written whenever ideas occurred and intended only for himself, including rudimentary intuitions that are sometimes noted rather clumsily, simply to pin down the thought as it

came to him. Colli calls the first kind of text exoteric and the second esoteric (1996, pp. 133 and 138). A number of experimental passages about the process of 'incorporation' belong to the second type, even though it is not possible to draw a firm dividing line.

2 See Gary Shapiro, *Nietzsche's Earth: Great Events, Great Politics* (Chicago/London: University of Chicago Press, 2016), p. 137: 'Nietzsche's most consistent name for this transformed earth is "garden".' See also my reflections on Shapiro's book in the Intermezzo (pp. 63–70).

3 See Christian Niemeyer (ed.), *Nietzsche-Lexikon* (Darmstadt: WBG, 2011), under 'Erde'.

4 I look further at 'knowledge' in Nietzsche's sense in Chapter 4.

5 English translation taken from Tyler T. Roberts, *Contesting Spirit. Nietzsche, Affirmation, Religion* (Princeton, NJ: Princeton University Press, 1998), p. 81.

6 See Shapiro, *Nietzsche's Earth*, chapter 5.

7 In Chapter 6 this image of humankind and its consequences for a contemporary philosophy of the earth are discussed in more detail.

8 Longinus J. Dohmen, *Nietzsche over de menselijke natuur. Een uiteenzetting van zijn verborgen anthropologie* (Kampen: Kok, 1994).

9 Gerrit Komrij, *Over de noodzaak van tuinieren*. Huizinga Lecture (Amsterdam: Bert Bakker, 1991).

10 'The most industrious people will find that it involves too much work simply to observe how differently men's instincts have grown, and might yet grow, depending on different moral climates. It would require whole generations, and generations of scholars who would collaborate systematically, to exhaust the points of view and the material. The same applies to the demonstration of the reasons for the differences between moral climates ("why is it that the sun of one fundamental moral judgment and main standard of value shines here and another one there?"). And it would be yet another job to determine the erroneousness of all these reasons and the whole nature of moral judgments to date' (GS, I, §7).

11 Patrick Wotling, *Nietzsche et le problème de la civilisation* (Paris: PUF, 1995), p. 270.

12 On Nietzsche's ideas about the European experiment, see Shapiro, *Nietzsche's Earth*, ch. 3, esp. p. 96.

13 It is therefore all the more galling that his sister Elisabeth Förster, who controlled his work after his death, had strong National Socialist sympathies. Nietzsche was fiercely opposed to such ideas, but Elisabeth placed his work in that light. She even made Hitler a present of Nietzsche's walking stick. Nietzsche's reception in the first half of the twentieth century, especially in Germany and Italy, was

very much influenced by this National Socialist interpretation. See also Paul van Tongeren, *Elementaire deeltjes – Nietzsche* (Amsterdam: AUP, 2016), esp. ch. 7.

14 Friedrich Kaulbach has defended the assertion that Nietzsche's idea of experimentation formed the organizational core of his entire philosophy; see his book *Nietzsches Idee Einer Experimentalphilosophie* (Vienna & Cologne: Böhlau, 1980). See also Volker Gerhardt, '"Experimental-Philosophie" Versuch einer Rekonstruktion' (1986) and Reinhart Maurer, 'Friedrich Kaulbach, Nietzsches Idee einer Experimentalphilosophie' (1983), as quoted in Niemeyer, *Nietzsche-Lexikon*, under 'Experimental Philosophie'.

Chapter 3

1 In his book *Grounding the Nietzsche Rhetoric of Earth* (Berlin & New York: De Gruyter, 2004), Adrian Del Caro analyses the rhetorical dimension of Nietzsche's language and its significance for his philosophy of the earth. 'I submit that the most serious use to which Nietzsche can be put . . . is the reclamation and the preservation of the earth. He made this his task, he set the standard at the threshold of the ecological age for humanity's first attempt to dwell affirmatively, intelligently and in partnership with the earth. What I refer to as partnership between humans and earth is expressed in the phrase "the superhuman is the meaning of the earth" found in the prologue to *Thus Spoke Zarathustra*. . . . Zarathustra is the conceptual ecosystem in which a superhuman could emerge' (pp. 49 and 51).

2 The quotations are taken from Friedrich Nietzsche, *Thus Spoke Zarathustra. A Book for All and None*, trans. Adrian Del Caro (Cambridge, UK: Cambridge University Press, 2006). For reasons of drama, they do not always occur in the order in which they are found in the book.

3 The word '*Übermensch*' had a major impact on Nietzsche's reception, especially in German and Italian publications prior to 1945. Even then, the figure was interpreted in many different ways. (See Niemeyer, *Nietzsche Lexikon*, under 'Übermensch'.) The term occurs only in *Thus Spoke Zarathustra* and in references to it in *Ecce Homo*, in letters and notebooks and so on. The overman's opposite number is 'the last human being, who makes everything small' (Z, I, 'Zarathustra's Prologue, §5') and who has learnt 'to perceive all diversity as immoral' (NF-1880, 3[98]). In my interpretation the overman is connected to Nietzsche's efforts to make the earth into the source and horizon of a new paradigm of epistemology. He breaks with 'modernity', which takes the human species as its source and horizon. See also Chapter 1, note 10.

4 Vesuvius is not on an island, but from Nietzsche's window in the house in Sorrento it appears as if it is. This is the explanation Paulo D'Iorio gives in his *Nietzsche's Journey to Sorrento* (Chicago & London: University of Chicago Press, 2016), ch. 4, 'Sorrentiner Papiere', esp. p. 155. Another explanation is that the sea has a symbolic significance here and refers to boundlessness and infinity.

5 Zarathustra may be 'transcribing' here a certain scientific representation of organic life, but he offers no further explanation. In his posthumously published writings from this time we see that Nietzsche was reading books about biology and evolution. Was this passage perhaps partly inspired by what he had read in them? The growth of organisms was analysed as an interplay of forces that order and obey each other, and grow in the direction of higher and higher complexity of form. One explanation given for this was that the cells obey orders from a higher authority as the organism is constructed. Might it be that Nietzsche picked up this idea from the life sciences and included it in his vision of the earth? (And does it have something to do with what contemporary biologists call the autopoiesis of the organism?)

6 While still very young, in his master's thesis, Nietzsche had criticized Kant's concept of autonomy. Kant believed he could evaluate good and evil based on pure reason, but young Nietzsche opposed this view from the perspective of the organic.

7 The eagle and the snake accompany Zarathustra at all times. The eagle, king of the birds, in 'the tree called future', is the first to see, with his clear and sharp eyes, the sun and the new dawn. The snake, 'the wisest animal under the sun', symbolizes among other things the circularity of time and the eternal return. Nietzsche was intending to explain the biblical significance of the snake as a symbol of evil, connected with female cunning and temptation. Both in the Bible and in Greek and Roman writing, the eagle and the snake are loaded symbols, and Nietzsche was very familiar with them. See Niemeyer, *Nietzsche-Lexikon* under 'Adler' and 'Schlange'.

8 van Tongeren, *Elementaire deeltjes – Nietzsche*. See ch. 4, in which he explains the sense in which Z can be regarded as a performance of failure (p. 149).

Chapter 4

1 I am especially grateful in this respect to Atem and Oesha Ramsundersingh, Lee Hong Yuan, Christopher Chuang, Jan Willem Kirpestein, Herman and Herma Wijffels, Mary Evelien Tucker, Irene Von Lippe, Harry Kunneman, Fernando Suarez Muller, Caroline Suransky, and the team of lecturers and students at the

international Summer School of the University of Humanistic Studies in Utrecht, who have been my interlocutors for more than ten years.

2   As well as Nietzsche's work, I read the letters of Vincent van Gogh, an impressive account of his discovery of the earth in the south of France, brilliantly illustrated in the six-volume *Vincent van Gogh – The Letters. The Complete Illustrated and Annotated Edition*, ed. Leo Jansen, Hans Luijten and Nienke Bakker (London: Thames and Hudson, 2009). The digital edition of the letters is available at http://vangoghletters.org/vg/letters.html. See also Sylviane Bonte and Yves Séméria, *Nietzsche et van Gogh. Incandescences maudites* (Nice: Ovadia, 2012).

3   'Gaia' is the name that climatologist James Lovelock introduced in 1979 to refer to the earth, when he was defending his idea that the earth is not a heavenly body among others but a complex living entity made up of a web of ecosystems that are interconnected and influence each other. I look at this further in Chapter 6.

4   Stephen Harding, *Animate Earth. Science, Intuition and Gaia*, 2nd edn (White River Junction, VT: Chelsea Green Books, 2009).

5   Bernard Stiegler, *Disruption. Comment ne pas devenir fou* (Paris: LLL, 2016). French science fiction writer Alain Damasio attempts to expand upon this idea in his novels and stories. His books are intended to provide shock therapy to combat these developments and to introduce new ways of putting up resistance, new means of flight and 'creative piracy'.

6   All the information and much more besides can be found on the excellent website www.Footprintnetwork.org

7   Harald Welzer, *Klimakriege. Wofür im 21. Jahrhundert getötet wird* (Frankfurt am Main: Fischer, 2008).

8   For my explanation of Nietzsche's interpretation of Epicurus and the Stoics I have made extensive use of what Dorian Astor has written on the subject in his highly informative book *Nietzsche. La Détresse du présent* (Paris: Gallimard, 2014), esp. ch. 9, 'Micropolitique de l'éternité'.

9   Astor, *Nietzsche*, p. 512. See also GS §306: '*Stoics and Epicureans.* The Epicurean selects the situation, the persons, and even the events that suit his extremely irritable, intellectual constitution; he gives up all others, which means almost everything, because they would be too strong and heavy for him to digest.'

10 Another fragment from that same year contains a similar approach: 'To proceed from the smallest nearest . . .' (*Vom Kleinsten Nächsten auszugehen*) (NF-1881, 13[20]).

11 On Gandhi and ecology, see Saskia van Goelst Meijer, *Profound Revolution. Towards an Integrated Understanding of Contemporary Nonviolence* (Utrecht: University of Humanistic Studies, dissertation, 2015), esp. ch. 4.

12 Philippe Granarolo has devoted an entire book to this aspect of Nietzsche's philosophy, entitled *Nietzsche. Cinq scénarios pour le futur* (Paris: Editions Les Belles Lettres, 2014).
13 UN report *Our Common Future* (1987), ch. 2: 'Towards Sustainable Development'.
14 See Axel Gosseries and Lukas H. Meyer (eds.), *Intergenerational Justice* (Oxford: OUP, 2009). Carolina Suransky and J. C. van der Merwe write in 'Transcending Apartheid in Higher Education. Transforming an institutional culture', in *Race Ethnicity and Education* 2015: 'Grosseries and Meyer demonstrate that the idea of linking intergenerational perspectives to justice claims has rapidly gained significance over the last few decades and has generated a discourse on "intergenerational justice". This concept has become particularly powerful in the field of sustainable development where "intergenerational duties" play an important role in addressing environmental concerns. In the world of sustainable development, one is acutely aware of intergenerational obligations and that "just contemplating the present does not suffice if long term sustainability is our goal".'

Chapter 5

1 UNESCO Universal Declaration on Cultural Diversity (2001)© http://portal.unesco.org/en/ev.php-URL_ID=13179&URL_DO=DO_TOPIC&URL_SECTION=201.html
2 Shapiro, *Nietzsche's Earth*, pp. 150 and 151.
3 In Chapter 2 I showed how Nietzsche was influenced by Burckhardt's reflections on gardens.
4 French philosophers Gilles Deleuze and Félix Quatarri introduced the terms 'geo-aesthetics' and 'geo-politics'. Shapiro regularly refers to their thinking and builds upon it.
5 Richard Evanoff, *Bioregionalism and Global Ethics* (New York: Routledge, 2010), p. 152.
6 According to Patrick Wotling, the notion of '*Kultur*/culture' forms 'the organizational centre' of Nietzsche's thought, which itself underlines its 'foundational status'. Wotling writes that the composition of a typology of cultures, conceived as complex collective entities, distinguished both geographically and historically, forms one of the permanent points of special interest of Nietzsche's thought. See Dorian Astor, *Dictionnaire Nietzsche* (Paris: Robert Laffont, 2017) under 'Culture' (pp. 209–13) and Céline Denat and Patrick

Wotling, *Dictionnaire Nietzsche* (Paris: Ellipses, 2013) under 'Culture (*Cultur/ Kultur*)' (pp. 95–101). (NB. Wotling wrote the entry for Culture in both Astor's *Dictionnaire* and his own.)

7 *Indigenous and Traditional Peoples of the World and Ecoregion Conservation: An Integrated Approach to Conserving the World of Biological and Cultural Diversity* (WWF International & Terralingua, Gland, Switzerland, 2000).

8 Pascal Picq, *De Darwin à Lévi-Strauss. L'homme et la diversité en danger* (Paris: Odile Jacob, 2013).

9 To anyone wanting to know more on the subject I can recommend the book *Cultural and Spiritual Values of Biodiversity: A Complementary Contribution to the Global Biodiversity Assessment* (London: UNDP, 2000), a publication of the United Nations Development Programme (UNEP). It includes articles about very diverse indigenous cultures, explaining their cosmologies, illustrated with an impressive arsenal of examples.

10 Philippe Descola, *Diversité naturelle, Diversité culturelle* (Montrouge: Bayard, 2010).

11 John King Fairbank and Merle Goldman, *China: A New History. Second, Enlarged Edition* (Cambridge, MA: Harvard University Press, 2006), p. 64.

12 The catalogue for the exhibition Philippe Descola curated in 2011 in the Musée du Quai Branly in Paris is interesting in this connection: *La Fabrique des Images. Visions du monde et Formes de la Représentation* (Paris: Somogy éditions d'art, 3 February 2010).

13 James Clifford, *Returns. Becoming Indigenous in the Twenty-first Century* (Cambridge MA: Harvard University Press, 2013).

14 A good example of this is the opening sentence of the KARI-OCA 2 Declaration: 'We, the Indigenous Peoples of Mother Earth assembled at the site of Kari-Oka I, sacred Kari-Oka Púku, Rio de Janeiro to participate in the United Nations Conference on Sustainable Development Rio+20,' Indigenous Peoples Global Conference on Rio+20 and Mother Earth. http://www.ienearth.org/. Rio+20 is the name of the international conference on sustainable development organized by the United Nations twenty years after the first conference in 1992.

15 See also the photobook *Before They Pass Away* (Kempen: Neues Verlag, 2013), with photographs by Jimmy Nelson of a hundred still extant local cultures.

16 I will return to this in Chapter 6.

17 Epeli Hau'ofa was professor of ethnology at the University of the Pacific on the Fiji islands, where he founded the Oceanic Centre for Arts and Culture. See Clifford, *Returns*, p. 195.

18 Hau'ofa (2008: 67); cited in Clifford, *Returns*, p. 42.

19 See ch. 3, 'Act Three' of *Thus Spoke Zarathustra*.

20 UNESCO's Ecological Sciences for Sustainable development: 'Agricultural practices actually determine the way human beings, be they from cities or rural areas, relate to their environment, consume, and manage their biological and cultural heritage.... To face tomorrow's challenges, we will consequently need to develop a new "agro-culture" which maintains diversity and achieves sustainability.' www.Unesco.org.

21 'The existing food system has failed and needs urgent reform, according to UN expert Olivier de Schutter who argues there should be a greater emphasis on local food production. Rebuilding local food systems, for instance, would decentralize food systems, making them more flexible and creating links between cities and rural hinterlands. We now recognize that poor, food-deficit countries should be supported not by trade and aid alone, but first and foremost by supporting them in their ability to feed themselves.' http://www.theguardian.com/ From 2008 to 2013 Olivier de Schutter was United Nations Special Rapporteur on the right to food. In his closing report he made far-reaching recommendations concerning how to reform and strengthen local agriculture.

22 Vandana Shiva, *Soil Not Oil. Environmental Justice in an Age of Climate Crisis* (Berkeley, CA: North Atlantic Books, 2008), *Staying Alive. Women, Ecology, and Development* (Berkeley, CA: North Atlantic Books, 2010) and *Making Peace with the Earth* (London: Pluto Press, 2013). The work of Indian social scientist Ashis Nandy has also had a major influence on this development. See *A Very Popular Exile* (Oxford: OUP, 2007) as well as *Talking India: Ashis Nandy in Conversation with Ramin Jahanbegloo* (Oxford: OUP, 2006).

23 See Eduardo Gudynas, 'Buen Vivir: Today's Tomorrow', *Development* 54, no. 4 (2011): 441–7.

24 Cross-Cultural Foundation of Uganda, *Culture in Governance: Does It Work?* (Kampala: CCFU, 2010); Cross-Cultural Foundation of Uganda, *The Family at the Heart of Managing Cultural Diversity. Conversations with 35 Ugandan Leaders and Rural Women and Men* (Kampala: CCFU, 2011); Emily Drani, Santa I. Kayonga and John de Coninck, *Kosmopolis Institute, University of Humanistic Studies, Pluralism Working Paper Series no 8.* (2014) *Managing Cultural Conflict in the Rwensori Region: Interventions and Aspirations*.

25 The 'Commons' movement has also had a part to play. See www.onthecommons.org.

26 'The term biocultural region designates a local geographic area in which specific human cultures develop in relation to the natural ecosystems they inhabit.' Evanoff, *Bioregionalism and Global Ethics*.

27 Mark Hathaway and Leonardo Boff, *The Tao of Liberation. Exploring the Ecology of Transformation* (New York: Orbis, 2010), p. 11.

28 www.GlobalFootprintNetwork.Org.
29 Edward Glaeser, *Triumph of the City* (London: Penguin, 2011).
30 John Elkington, *The Zeronauts. Breaking the Sustainability Barrier* (London: Routledge, 2012).
31 Nietzsche's criticism of the spiritualization of morality and his plea for a more bodily and earthbound vision of life was taken up by a number of philosophers several decades later. French thinkers in particular, including Georges Bataille, Michel Foucault, Gilles Deleuze, Félix Guattari and Luce Irigaray, brought about a revival of this aspect of Nietzschean thinking in Paris in the twentieth century, provoking responses worldwide. It is striking however that the first three of these thinkers did not generally explicitly link the turn to a more physical way of philosophizing with the perspective of the earth. Guattari and above all Irigaray did do so. It is my conviction that Nietzsche's injunction to philosophize according to the 'guide of the body' attains its full significance only in the context of the relationship between humans and the earth and the transformation into the overman. The rediscovery of how the local and planetary dimensions are linked is part of such a turnaround.
32 Tu Weiming, 'The Ecological Turn in New Confucian Humanism', in Tu Weiming and Mary Evelyn Tucker (eds.), *Confucian Spirituality*, vol. 2 (New York: Crossroad, 2004). See also Tu Weiming, 'The Continuity of Being: Chinese Visions of Nature', in Mary Evelyn Tucker and John Berthrong (eds.), *Confucianism and Ecology* (Cambridge, MA: Harvard University Press, 1998).
33 I have looked in more detail at how this vision can inspire Western humanism in Henk Manschot, 'Leven op aarde. Het verhaal van de mens', in Hans Alma and Adri Smaling (eds.), *Waarvoor je leeft. Studies naar humanistische bronnen van zin* (Amsterdam: SWP, 2009), pp. 59–85.
34 See H. A. M. Manschot and A. C. Suransky, 'From a Human-centered to a Life-centered Humanism: Three Dimensions of an Ecological Turn', in D. McGowen and A. B. Pinn (eds.), *Everyday Humanism* (London: Equinox, 2014), pp. 125–37.
35 Ezio Manzini, *Design. When Everybody Designs. An Introduction to Designs for Social Innovation* (Cambridge, MA: MIT Press, 2015), p. 25.
36 Arturo Escobar, *Designs for the Pluriverse. Radical Interdependence, Autonomy, and the Making of Worlds* (Durham, NC: Duke University Press, 2018), p. 208.
37 http://www.transitiontowns.nl and https://transitionnetwork.org.
38 http://colibris-lemouvement.org. The Colibri movement was established by Pierre Rahbi. An Algerian by origin, he lives in France, where the movement is extremely influential. Rahbi is the author of books including *Manifeste pour la Terre et l'Humanisme* (Arles: Actes Sud, 2008).
39 See Escobar, *Designs for the Pluriverse*.

40   United Nations, *Transforming our World – The 2030 Agenda for Sustainable Development*. See https://sustainabledevelopment.un.org/post2015/transformingourworld.

41   Let us briefly compare this cultural approach to that produced by abstract universal values, the strategy that took as its starting point the values of freedom, equality and fraternity, and developed these into the rights of man, applicable to all. Since then the question has repeatedly arisen as to how these universal rights can be given shape in specific local contexts. The local appears as a particularization of the universal. Communities are called upon to give local content to human rights. The problems arising as a result were obviated by the addition of special rights for cultural minorities. Then came special conventions, which recognized the rights of indigenous cultures and surrounded their particular cultural patterns, customs and traditions with protective structures. A simple list of their titles illustrates this development: *The Convention on Biological Diversity* (1993), *The Unesco Declaration on Cultural Diversity* (2003), *The Declaration of the Rights of Indigenous People* (2007). I cannot escape the impression that the conceptual framework itself – from universal to specific – lies at the heart of the universal approach and determines its dynamic, causing increasing friction. Indian professor S. Kakarala, who has researched this development in the field of human rights, concludes, 'Diversity has shifted from a category between others to a foundational value of human life and humanity. This implies a fundamental shift, epistemologically as well as ontologically. It implies that all central categories of being human, such as liberty, equality, autonomy and agency have to be reframed and to be mediated through the categories of bio- and cultural diversity.' – S. Kakarala, 'Key-topics in Post-colonial Studies' (unpublished manuscript). Comparing top-down and bottom-up approaches raises the question of whether a Nietzschean-inspired approach might provide an alternative, or at least a fruitful supplement.

42   By giving the local an ecological orientation, Nietzsche was able to nip in the bud any nationalist or fascist interpretation of the concept. See ch. 2, 'Cosmos and climate'.

Chapter 6

1   Spacelog Apollo 8 transcripts, via https://apollo8.spacelog.org/page/03:13:46:23/
2   Archibald MacLeish, 'Riders on Earth Together, Brothers in Eternal Cold', *The New York Times*, 25 December 1968.

3 Quoted by David Beaver in 'The Case for Planetary Awareness. How the New Space Age will Profoundly Change Our Worldview', via http://overview-effect.earth/ First published in *Space Times, the Magazine of the American Astronautical Society*, vol. 55, March/April 2016.
4 Quoted in James Gunn, *Alternate Worlds: The Illustrated History of Science Fiction*, 3d edn (Jefferson, NC: McFarland and Company, 2018), p. 220.
5 See www.theovervieweffect.org.
6 Wilhelm Schmid, *Ökologische Lebenskunst* (Berlin: Suhrkamp, 2008), ch. 1.
7 Paul Crutzen and Eugene Stoermer, 'The Anthropocene', International Geosphere–Biosphere Program, *Global Change Newsletter*, no. 41 (2000): 17; Wil Steffen, Jacques Grinevald, Paul Crutzen and John MacNeill, 'Conceptual and Historical Perspectives', in *Philosophical Transactions of the Royal Society. A* (2011): 842–67.
8 An additional merit of the concept of the Anthropocene, the researchers argue, is that human dealings are included in the science of system earth. Humans too are from now on the object of study in the geo- and biosciences. The Anthropocene therefore calls into question the dividing line between the natural sciences and the social sciences. Since the introduction of the concept, a broad discussion has arisen about its definition and reach. An outline of this discussion can be found in Christophe Bonneuil and Jean-Baptiste Fressoz, *L'Evénement Anthropocène. La Terre, l'Histoire et Nous* (Paris: Du Seuil, 2016).
9 Johan Rockström et al., 'A Safe Operating Space for Humanity', *Nature*, no. 461 (24 September 2009). It was Rockström who put forward the idea that there are planetary boundaries that have already been crossed.
10 Bruno Latour, 'Waiting for Gaia: Composing the Common World through Arts and Politics'. A lecture at the French Institute, London, November 2011.
11 Isabelle Stengers, 'Penser á partir du ravage écologique', in *De l'univers clos au monde infini*, ed. Emilie Hache (Bellevaux: Dehors, 2014).
12 Bruno Latour, *Politics of Nature. How to Bring Sciences into Democracy* (Cambridge, MA: Harvard University Press, 1999/2004). Bruno Latour, *Face á Gaia. Huits Conférences sur le nouveau régime climatique* (Paris: La Découverte, 2015).
13 Peter Sloterdijk, *Du musst dein Leben ändern. Über Anthropotechnik* (Frankfurt am Main: Suhrkamp, 2009).
14 Lynn Margulis, *Symbiotic Planet: A New Look at Evolution* (New York: Basic Books, 1998).
15 It is interesting to read the classical traditions of cosmopolitanism and world citizenship afresh through the lens of Nietzsche's philosophy of the earth. The local–planetary paradigm might throw new light on the ancient question of how

people can be faithful, simultaneously, to the local community of which they are part and to all human beings. Martha Nussbaum, who has put cosmopolitanism back on the philosophical agenda, recognizes the tension here but explains it as a tension between universal values and cultural diversity. Kwame Appiah has introduced the useful term 'rooted cosmopolitanism'. He points rather more emphatically to the importance of everyone's cultural baggage and identity in dealing with contemporary issues surrounding world citizenship. But neither takes the earth as the framework for their philosophy. In future research I want to look more closely at the subjects of world citizenship and cosmopolitanism as seen from a Nietzschean local–planetary perspective.

16 Not to be confused with the naturalism of Philippe Descola, who uses the term to refer to the objectification of nature in Western thinking (see Chapter 5).

17 In *Beyond Good and Evil* Nietzsche uses the expression '*Grundtext Homo Natura*' (BGE, §230). Del Caro comments: 'Nietzsche refers in this passage to the basic text of *homo natura* as being terrible (*Schrecklich*). . . . The task of philosophers is to translate human back into nature, master the interpretations and connotations that have been scribbled over the eternal *Grundtext Homo Natura*, make it so that humans stand before their own nature as they stand before the rest of nature, intrepidly, not seduced by the flattery of metaphysics. This may be an insane task, but it remains a task.' Del Caro, *Grounding the Nietzsche Rhetoric of Earth*, pp. 21–2.

18 Frans de Waal has provided us with a different vision. For instance, in his book *The Bonobo and the Atheist. In Search of Humanism among the Primates* (New York: Norton, 2013), he defends the idea that moral attitudes are already present in primates and that humans further build upon them.

19 *Thus Spoke Zarathustra* opens with the scene of a man dancing on 'a rope fastened between animal and overman – a rope over an abyss'.

20 See Deborah Danowski and Eduardo Viveiros de Castro, 'L'Arrêt de Monde', in *De l'Univers Clos au Monde Infini. Textes réunis et présentes par Emilie Hache* (Arle: Dehors, 2014).

21 See for example Isabelle Stengers in 'Penser à partir du ravage écologique' and Clive Hamilton, *Requiem for a Species. Why We Resist the Truth about Climate Change* (London: Earthscan, 2010), ch. 8.

22 Bruno Latour, 'Steps Toward the Writing of a Compositionist Manifesto', *New Literary History* 41 (2010): 471–90.

23 In the ancient Greek myths and legends, the Argonauts are the crew of the ship Argo (built by Argos) on which Greek hero Jason goes in search of the Golden Fleece. Jason's sea voyage is at least as much of an adventure as that of Odysseus. The legend is even older than Homer's *The Iliad* and *The Odyssey*.

Afterword

1 *Mittag und Ewigkeit. Fingerzeige zu einem neuen Leben* (NF-1881, 11[195]).
2 *Comment tout peut s'effondrer. Petit manuel de collapsologie à l'usage des générations présentes* (Paris: Du Seuil, 2015); *Une autre fin du monde est possible. Vivre l'effondrement (et pas seulement survivre)* (Paris: Du Seuil, 2018).
3 See Chapter 5 for the problem of urbanization.
4 Manzini, *Design. When Everybody Designs*, p. 25.
5 Escobar, *Designs for the Pluriverse*, p. 208.
6 Manzini, 'Resilient Systems and Cosmopolitan Localism. The Emerging Scenario of the Small, Local, Open and Connected Space', in *Economy of Sufficiency. Essays on Wealth in Diversity, Enjoyable Limits and Creating Commons*, ed. Uwe Schneidewind, Tilman Santarius and Anja Humburg (Wuppertal Institut für Klima, Umwelt, Energie GmbH, 2013).

Bibliography

A

Acampora, Christa Davis and Ralph Acampora (eds.). *A Nietzschean Bestiary. Becoming Animal beyond Docile and Brutal*. Washington, DC: Rowman and Littlefield, 2004.

Actuel Marx No. 56 (2014). *Les Amériques Indiennes face au Néo-libéralisme*. Paris: Presse Universitaire de France.

Afeissa, Hicham-Stéphane (ed.). *Ethique de l'Environnement. Nature, Valeur, Respect. Textes Réunis*. Paris: Librairie Philosophique Vrin, 2007.

Afeissa, Hicham-Stéphane. *Philosophie animale. Différence, responsabilité et communauté. Textes réunis*. Paris: Librairie Philosophique Vrin, 2010.

Anbeek, C. W. and H. A. Alma. 'Worldviewing Competence for Narrative Interreligious Dialogue. A Humanist Contribution to Spiritual Care' (2013), in D. S. Shipani (ed.), *Multifaith Views in Spiritual Care*, pp. 149–69. Kitchener, ON: Pandora Press, 2014.

Appadurai, Arjun. *The Future as Cultural Fact. Essays on the Global Condition*. New York/London: Verso, 2013.

Appiah, Kwame Anthony. *Cosmopolitanism. Ethics in a World of Strangers*. New York/London: W.W. Norton/Allen Lane, 2006.

Astor, Dorian (ed.). *Dictionnaire Nietzsche*. Paris: Robert Laffont, 2017.

Astor, Dorian. *Nietzsche. La détresse du présent*. Paris: Gallimard, 2014.

B

Blondel, Eric. *Nietzsche, le corps et la culture*. Paris: Presse Universitaire de France, 1986.

Bonneuil, Christophe and Jean-Baptiste Fressoz. *L'Evénement Anthropocène. La terre, l'histoire et nous*. Paris: Éditions du Seuil, 2016.

Bonneuil, Christophe and Jean-Baptiste Fressoz. *The Shock of the Anthropocene*. New York/London: Verso, 2016.

Bonte, Sylviane and Yves Séméria. *Nietzsche – Van Gogh. Incandescences maudites*. Nice: Les Éditions Ovadia, 2012.

Bouriau, Christophe. *Nietzsche et la Renaissance*. Paris: Presse Universitaire de France, 2015.

Brémondy, François. *Bestiaire de Friedrich Nietzsche*. Mons: Les Éditions Sils Maria, 2011.

Bulhof-Rutgers, I. N. *Apollos Wiederkehr. Eine Untersuchung der Rolle des Kreises in Nietzsches Denken. Über Geschichte und Zeit*. The Hague: Martinus Nijhof, 1969.

C

Chakrabarty, Dipesh. *Provincializing Europe. Postcolonial thought and Historical Difference*. With a new preface by the author. Princeton NJ: Princeton University Press, 2000.

Clifford, James. *Returns. Becoming Indigenous in the Twenty-first Century*. Cambridge, MA: Harvard University Press, 2013.

Colli, Giorgio. *Écrits sur Nietzsche*. Trans. Patricia Farazzi. Paris: Éditions de l'Éclat, 1996.

Colli, Giorgio. *Scritti su Nietzsche*. Milan: Adelphi, 1980.

Conche, Marcel. *Nietzsche et le bouddhisme*. Paris: Encre Marine, 1997.

Connolly, William E. *Facing the Planetary. Entangled Humanism and the Politics of Swarming*. Durham, NC: Duke University Press, 2017.

Coolen, Martin and Koo van de Wal. *Het eigen gewicht der dingen*. Eindhoven: Uitgeverij Damon, 2002.

Cross-Cultural Foundation of Uganda (CCFU). *Culture in Governance. Does It Work?* Kampala: CCFU, 2010.

Cross-Cultural Foundation of Uganda (CCFU). *The Family: At the Heart of Managing Cultural Diversity. Conversations with 35 Ugandan Leaders and Rural Women and Men*. Kampala: CCFU, 2011.

Crutzen, Paul and Eugene Stoermer. 'The Anthropocene'. International Geosphere–Biosphere Programme', *Global Change Newsletter*, no. 41 (2000): 17.

D

Damasio, Alain. *La zone du dehors*. Paris: Gallimard, Folio, 1997/2015.

Damasio, Alain. *La horde du Contrevent*. Paris: Gallimard, Folio, 2014.

Danowski, Deborah and Eduardo Viveiros de Castro. 'L'Arrêt de Monde', in *De l'Univers Clos au Monde Infini. Textes réunis et présentés par Emilie Hache*. Bellevaux: Éditions Dehors, 2014.

Delbard, Olivier. *Par-delà le développement Durable. Nature, Culture(s) et Coopération(s)*. Lormont: Le Bord de l'Eau, 2014.

Del Caro, Adrian. *Grounding the Nietzsche Rhetoric of Earth*. Berlin/New York: De Gruyter, 2004.

Deleuze, Gilles. *Nietzsche et la Philosophie*. Paris: Presse Universitaire de France, 1970.

Denat, Céline and Patrick Wotling. *Dictionnaire Nietzsche*. Paris: Ellipses Marketing, 2013.

Descola, Philippe. *Par-delà nature et culture*. Paris: Gallimard, 2005.

Descola, Philippe. *Diversité des natures, diversité des cultures*. Montrouge: Bayard, 2010.

Descola, Philippe. *La Fabrique des Images. Visions du monde et formes de la représentation*. Paris: Musée du Quai Branly, 2011.

Dohmen, Longinus [Joep]. *Nietzsche over de menselijke natuur. Een uiteenzetting van zijn verborgen antropologie*. Kampen: Kok, 1994.

Dohmen, Joep (ed.). *Over Levenskunst. De grote filosofen over het goede leven*. Amsterdam: Ambo, 2002/2014.

Dohmen, Joep. 'Philosophers on the "Art-of-Living"', *Journal of Happiness Studies* 4 (December 2003): 351–71.

Drenthen, Martin. 'Grenzen aan Wildheid. Wildernisverlangen en de betekenis van Nietzsches moraalkritiek voor de actuele milieuethiek' (Nijmegen dissertation). Eindhoven: Damon, 2003.

Duyndam, Joachim, Marcel Poorthuis and Theo de Wit (eds.). *Humanisme en Religie. Controversen, bruggen, perspectieven*. Delft: Eburon, 2005.

E

Van Egmond, Klaas. *Een vorm van beschaving*. Zeist: Christofoor, 2010.

Escobar, Arturo. *Designs for the Pluriverse. Radical Interdependence, Autonomy, and the Making of Worlds*. Durham, NC: Duke University Press, 2017.

Evanoff, Richard. *Bioregionalism and Global Ethics. A Transactional Approach to Achieving Ecological Sustainability, Social Justice, and Human Well-being*. New York: Routledge, 2010.

F

Foucault, Michel. *Les mots et les choses. Une archéologie des sciences humaines*. Paris: Gallimard, 1966.

Foucault, Michel. *Histoire de la sexualité. I, II, III*. Paris: Gallimard, 1984.

Foucault, Michel. *Ervaring en waarheid*. (Duccio Trombadori in conversation with Michel Foucault). Nijmegen: SUN, 1985.

Foucault, Michel. *Le Gouvernement de soi et des autres. Cours au Collège de France, 1982–1983*. Paris: EHESS, Gallimard/Seuil, 2008.

Fresco, Louise O. *Hamburgers in het paradijs. Voedsel in tijden van schaarste en overvloed*. Amsterdam: Bert Bakker, 2012.

Fresco, Louise O. *Hamburgers in Paradise. The Stories behind the Food We Eat*. Trans. Liz Waters. Princeton and Oxford: Princeton University Press, 2016.

Fresco, Louise O. *Kruisbestuiving. Over kennis, kunst en het leven*. Amsterdam: Prometheus/Bert Bakker, 2014.

G

Gandhi, M. K. *An Autobiography. Or the Story of My Experiments with Truth*. Ahmedabad: Navajivan Publishing House, 1927.

Glaeser, Edward. *Triumph of the City. How Our Greatest Invention Makes Us Richer, Smarter, Greener, Healthier and Happier*. Harmondsworth, UK: Penguin, 2011.

Gosseries, Axel and Lukas H. Meyer (eds.). *Intergenerational Justice*. Oxford: Oxford University Press, 2009.

Granarolo, Philippe. *L'individu éternel. L'expérience Nietzschéenne de l'éternité*. Paris: Librairie Philosophique Vrin, 1993.

Granarolo, Philippe. *Nietzsche et les voies du Surhumain*. Nice: Canopé – Crdp 06, 2012.

Granarolo, Philippe. *Nietzsche. Cinq scénarios pour le futur*. Paris: Les Belles Lettres, 2014.

Grim, John and Mary Evelyn Tucker. *Ecology and Religion*. Washington: Island Press, 2014.

H

Hache, Emilie (ed.). *Ecologie Politique. Cosmos, communautés, milieu*. Paris: Éditions Amsterdam, 2012.

Hache, Emilie (ed.). *De l'Univers clos au monde infini, Textes réunis et présentes par Emilie Hache*. (Bruno Latour, Christophe Bonneuil, Pierre de Jouvancourt, Dipesh Chakrabarty, Isabelle Stengers, Giovanna Di Chiro, Déborah Danowski and Eduardo Viveiros de Castro.) Bellevaux: Éditions Dehors, 2014.

Hadot, Pierre. 'Réflexions sur la notion de "culture de soi"', in *Michel Foucault philosophe, Rencontre internationale, Paris 9, 10, 11 janvier 1988*, pp. 261–9. Paris: Éditions du Seuil, 1989.

Hadot, Pierre. *La Citadelle Intérieure. Introduction aux Pensées de Marc Aurele*. Paris: Fayard, 1997.
Hadot, Pierre. *La philosophie comme manière de vivre. Entretiens avec Jeannie Carlier et Arnold I. Davidson*. Paris: Albin Michel, 2001.
Hadot, Pierre. *Le Voile d'Isis. Essai sur l'histoire de l'idée de la nature*. Paris: Gallimard, 2004.
Hamilton, Clive. *Requiem for a Species. Why We Resist the Truth about Climate Change*. Washington/London: Earthscan, 2010.
Harding, Stephan. *Animate Earth. Science, Intuition and Gaia*. (Second Edition). White River Junction, VT: Chelsea Green Books, 2009.
Hathaway, Mark and Leonardo Boff. *The Tao of Liberation. Exploring the Ecology of Transformation*. New York: Orbis Books, 2009.
Hawken, Paul. *Blessed Unrest. How the Largest Movement in the World Came into Being and Why No One Saw It Coming*. New York: Viking, 2007.
Héber-Suffrin, Pierre. *Lecture d'Ainsi parlait Zarathoustra III: Penser, vouloir et dire l'éternel retour*. Paris: Éditions Kimé, 2012.
Hedlund-de Witt, Annick. 'Worldviews and the Transformation to Sustainable Societies. An Exploration of the Cultural and Psychological Dimensions of Our Global Environmental Challenges' (dissertation). Amsterdam: VU University, 2013.
Hefferman, George. 'A Life According to Nature. From Ancient Theoretical Ideal to Future Sustainable Practice' (conference paper, German consulate in New York 2010), *The Environmentalist*, no. 32, pp. 278–88. New York: Springer, 2012.

I

Iorio, Paulo D'. *Le Voyage de Nietzsche á Sorrente*. Paris: CNRS Éditions, 2012.
Iorio, Paolo D'. *Nietzsche's Journey to Sorrento. Genesis of the Philosophy of the Free Spirit*. Trans. Sylvia Mae Gorelick. Chicago and London: University of Chicago Press, 2016.
Irigaray, Luce. *Passions élémentaires*. Paris: Éditions de Minuit, 1982.
Irigaray, Luce. *Entre Orient et Occident*. Paris: Grasset, 1999.

J

Jonas, Hans. *The Imperative of Responsibility. In search of an Ethics for the Technological Age*. Chicago: University of Chicago Press, 1984.

K

Kant, Immanuel. 'Beantwortung der Frage: Was ist Aufklärung?', in *Berlinische Monatsschrift*. Berlin: Johann Erich Biester and Friedrich Gedike, 1784. English translation available at http://cnweb.cn.edu/kwheeler/documents/What_is_Enlightenment.pdf

Kant, Immanuel. *Zum Ewigen Frieden. Ein philosophischer Entwurf*. Königsberg: Friedrich Nicolovius, 1796.

Kant, Immanuel. *To Perpetual Peace. A Philosophical Sketch*. Trans. Ted Humphrey. Cambridge, MA: Hackett, 2003.

Komrij, Gerrit. *Over de noodzaak van tuinieren* (Huizinga Lecture). Amsterdam: Bert Bakker, 1991.

Kunneman, Harry. *Voorbij het dikke-ik. Bouwstenen voor een kritisch Humanisme*. Amsterdam: SWP, 2005.

Kunneman, Harry and Caroline Suransky. 'Cosmopolitanism and the Humanist Utopia', in Maria Rovisco and Magdalena Nowicka (eds.). *The Ashgate Research Companion to Cosmopolitanism*. Farnham: Ashgate, 2011.

Kuttner, H. G. *Nietzsche-Rezeption in Frankreich*. Essen: Die Baleue Eule, 1984.

L

Latour, Bruno. *Nous n'avons jamais été modernes. Essai d'anthropologie symétrique*. Paris: La Découverte, 1991.

Latour, Bruno. *Politics of Nature. How to Bring Sciences into Democracy*. Cambridge, MA: Harvard University Press, 1999/2004.

Latour, Bruno. *Cogitamus. Six Lettres sur les Humanités Scientifiques*. Paris: La Découverte, 2010.

Latour, Bruno. 'Steps Toward the Writing of a Compositionist Manifesto', *New Literary History*, 41 (2010): 471–90.

Latour, Bruno. 'Waiting for Gaia. Composing the Common World through Arts and Politics. A Lecture at the French Institute', in Albena Yaneva and Alejandro Zaera-Polo (eds.). *What Is Cosmopolitical Design?* pp. 21–33. Farnham: Ashgate, 2011.

Latour, Bruno. *Enquête sur les Modes de l'Existence. Une anthropologie des Modernes*. Paris: La Découverte, 2012.

Latour, Bruno. *Face à Gaia. Huits conférences sur le nouveau régime climatique*. Paris: La Découverte, 2015.

Latour, Bruno. *Facing Gaia. Eight lectures on the new climatic regime*. Trans. Catherine Porter. Cambridge, UK: Polity Press, 2017.

Latour, Bruno. *Où Atterrir? Comment s'orienter en politique?* Paris: La Découverte, 2017.
Latour, Bruno. *Down to Earth: Politics in the New Climatic Regime.* Cambridge, UK: Polity Press, 2018.
Lemaire, Ton. *Met open zinnen. Natuur, landschap, aarde.* Amsterdam: Ambo, 2002.
Lierop, Renske van. *Verdrongen kwetsbaarheid. Een theoretisch afstudeeronderzoek naar de duiding van grenservaringen*, 2014. Available via http://hdl.handle.net/11439/141
Light, Andrew and Holmes Rolston III (eds.). *Environmental Ethics. An Anthology.* Victoria: Blackwell, 2002.
Lippe-Biesterfeld, Irene van and Mathijs Schouten. *Leven in Verbinding.* Deventer: Ankh-Hermes, 2010.
Lourme, Louis. *Le nouvel âge de la citoyenneté mondiale.* Paris: Presse Universitaire de France, 2014.
Lovelock, James. *Gaia. A New Look at Life on Earth.* Oxford: Oxford University Press, 1979.
Lovelock, James. *The Revenge of Gaia. Why the Earth Is Fighting Back – and How We Can Still Save Humanity.* London: Penguin Books, 2006.

M

Manier, Bénédicte. *Un million de révolutions tranquilles. Travail, argent, habitat, santé, environnement... Comment les citoyens changent le monde.* Paris: Les Liens qui Libèrent, 2012.
Manschot, Henk. 'Nietzsche und die Postmoderne in der Philosophie', in Dietmar Kamper and Willen van Reijen (eds.). *Die unvollendete Vernunft: Moderne versus Postmoderne*, pp. 478–97. Frankfurt am Main: Suhrkamp, 1987.
Manschot, Henk. 'Moderne Gemeinschaft zwischen Warm und Kalt. Nietzsche und Kafka', in Zoltán Frenyó, Paul van Tongeren et al. (eds.). *Orientations. Papers Presented to the Conferences of Dutch and Hungarian Philosophers 1986-1990.* Maastricht: Shaker Publishing, 1987.
Manschot, Henk. 'Levenskunst of Lijfsbehoud? Een humanistische kritiek op het beginsel van autonomie in de gezondheidszorg'. Lecture in Utrecht: University of Humanistic Studies, 1992.
Manschot, Henk. 'Als wereldburger leven. Een probleemstelling', in Theo de Wit and Henk Manschot (eds.). *Solidariteit. Filosofische kritiek, ethiek en politiek,* pp. 194–213. Amsterdam: Boom, 1999.

Manschot, Henk. 'Human Rights and Human Diversity', in Annemie Halsema and Douwe van Houten. *Empowering Humanity. State of the Art in Humanistics*, pp. 177–93. Utrecht: De Tijdstroom, 2002.

Manschot, Henk, Jan-Willem Kirpestein and Vanno Jobse (eds.). *In Naam van de Natuur*. Kampen: Ten Have, 2009.

Manschot, Henk. 'Leven op aarde. Het verhaal van de mens', in Hans Alma and Adri Smaling (eds.). *Waarvoor je leeft. Studies naar humanistische bronnen van zin*, pp. 59–85. Amsterdam: SWP, 2010.

Manschot, Henk. 'Empowering Humanity. Ecologische kanttekeningen bij "de nieuwe menswetenschap"', in Hans Alma and Gerty Lensvelt-Mulder (eds.). *Waardevolle wetenschap. Zingeving en humanisering in het wetenschappelijk onderwijs*, pp. 35–45 Utrecht: University of Humanistic Studies, 2011.

Manschot, Henk and A. C. Suransky. 'From a Human-Centered to a Life-Centered Humanism, Three Dimensions of an Ecological Turn', in D. McGowen and A. B. Pinn (eds.). *Everyday Humanism*, pp. 125–37. London: Equinox, 2014.

Manzini, Ezio. *Design. When Everybody Designs. An Introduction to Designs for Social Innovation*. Cambridge, MA: MIT Press, 2015.

Margulis, Lynn. *Symbiotic Planet. A New Look at Evolution*. New York: Basic Books, 1998.

Margulis, Lynn. 'Gaia', in Emilie Hache (ed.). *Ecologie Politique*. Paris: Éditions Amsterdam, 2012.

Morin, Edgar. *Homeland Earth. A Manifesto for the New Millennium*, Creskill, NJ: Hampton Press, 1999.

N

Nandy, Ashis. *Talking India. Ashis Nandy in Conversation with Ramin Jahanbegloo*. Oxford: Oxford University Press, 2006.

Nandy, Ashis. *A Very Popular Exile*. Oxford/New Delhi: Oxford University Press, 2007.

Nelson, Jimmy. *Before They Pass Away*. (photobook). Kempen: Neues Verlag, 2013.

Niemeyer, Christian (ed.). *Nietzsche-Lexikon*. Darmstadt: WBG, 2011.

Nietzsche, *Penseur du chaos moderne*. Paris: Le Nouvel Observateur, 2007.

Nussbaum, Martha. *Cultivating Humanity. A Classical Defense of Reform in Liberal Education*. Cambridge, MA: Harvard University Press, 1997.

Nussbaum, Martha. *Creating Capabilities. The Human Development Approach*. Cambridge MA: Harvard University Press, 2011.

P

Parkes, Graham. 'Staying Loyal to the Earth: Nietzsche as an Ecological Thinker', in John Lippit (ed.). *Nietzsche's Futures*, pp. 167–88. Basingstoke: Macmillan, 1998.

Picq, Pascal. *De Darwin à Lévi-Strauss. L'Homme et la diversité en danger*. Paris: Odile Jacob, 2013.

Prideaux, Sue. *I am Dynamite. A Life of Friedrich Nietzsche*. London: Faber & Faber, 2018.

R

Rahbi, Pierre. *Manifeste pour la Terre et l'Humanisme. Pour une insurrection des consciences*. Arles: Actes Sud, 2008.

Ratner-Rosenhagen, Jennifer. *American Nietzsche. A History of an Icon and His Ideas*. Chicago/London: University of Chicago Press, 2012.

Renault, Matthieu. *L'Amérique de John Locke. L'expansion coloniale de la philosophie européenne*. Paris: Éditions Amsterdam, 2014.

Rockström, J. et al. 'A Safe Operating Space for Humanity', *Nature*, no. 461 (September 24, 2009).

S

Safranski, Rudiger. *Nietzsche: A Philosophical Biography*. Trans. Shelley Frisch. New York: W.W. Norton, 2002.

Schaeffer, Jean-Marie. *La fin de l'exception humaine*. Paris: Gallimard, 2007.

Schmid, Wilhelm. *Ökologische Lebenskunst. Was jeder Einzelne für das Leben auf dem Planeten tun kann*. Berlin: Suhrkamp, 2008.

Sen, Amartya. 'Democracy as a Universal Value', *Journal of Democracy* 10, no. 3 (1999): 3–17.

Shapiro, Gary. 'Nietzsche on Geophilosophy and Geoaesthetics', in Keith Ansell-Pearson (ed.). *The Blackwell Companion to Nietzsche*, pp. 477–94. New York: Blackwell, 2006.

Shapiro, Gary. 'Beyond Peoples and Fatherlands: Nietzsche's Geophilosophy and the Direction of the Earth', *Journal of Nietzsche Studies*, no. 35 (2008): 9–27.

Shapiro, Gary. *Nietzsche's Earth. Great events, Great Politics*. Chicago/London: University of Chicago Press, 2016.

Shiva, Vandana. *India Divided. Diversity and Democracy under Attack*. New York/ New Delhi: Seven Stories Press, 2005.

Shiva, Vandana. *Soil Not Oil. Environmental Justice in an Age of Climate Crisis*. Berkeley, CA: North Atlantic Books, 2015. (First published in Boston by South End Press, 2008).

Shiva, Vandana. *Staying Alive. Women, Ecology, and Development*. Berkeley, CA: North Atlantic Books, 2016. (First published in New Delhi by Women Unlimited, 1988/1999, and in Boston by South End Press, 2010).

Shiva, Vandana. *Making Peace with the Earth*. London: Pluto Press, 2012.

Sloterdijk, Peter. *Du musst dein Leben ändern. Über Anthropotechnik*. Frankfurt am Main: Suhrkamp, 2009.

Steffen, Wil, Jacques Grinevald, Paul Crutzen and John MacNeill. 'Conceptual and Historical Perspectives', *Philosophical Transactions of the Royal Society A*, no. 369 (2011): 842–67.

Stengers, Isabelle. 'The Cosmopolitical proposal / La proposition cosmopolitique', in Jacques Lolive and Olivier Soubeyran (eds.). *L'Emergence des cosmopolitiques*. Paris: La Découverte, 2007 [2004].

Stengers, Isabelle. *Cosmopolitics: Learning to Think with Sciences, Peoples and Nature*. Halifax NS: Keynote Address, Saint Mary's University, 2014.

Stengers, Isabelle. 'Penser á partir du ravage écologique', in Emilie Hache (ed.). *De l'univers clos au monde infini. Textes réunis et présentés par Emilie Hache*. Bellevaux: Éditions Dehors, 2014.

Stiegler, Barbara. *Nietzsche et la biologie*. Paris: Presse Universitaire de France, 2001.

Stiegler, Bernard. *Dans la disruption. Comment ne pas devenir fou*. Paris: Les Liens qui Libèrent, 2016.

T

Tongeren, Paul van. *Nietzsche als arts van de cultuur*. Kampen: Kok Agora, 1990.

Tongeren, Paul van. *Reinterpreting Modern Culture. An Introduction to Friedrich Nietzsche's Philosophy*. West Lafayette, IN: Purdue University Press, 2000.

Tongeren, Paul van. *CD boxed set Nietzsche en het nihilisme*. Leusden: ISVW Uitgevers, 2012.

Tongeren, Paul van. *Elementaire deeltjes – Nietzsche*. Amsterdam: Amsterdam University Press, 2016.

Tongeren, Paul van. *Friedrich Nietzsche and European Nihilism*. Newcastle upon Tyne: Cambridge Scholars Publishing, 2018.

U

United Nations Development Program. *Cultural and Spiritual Values of Biodiversity. A Complementary Contribution to the Global Biodiversity Assessment*, London: UNDP, 2000.

V

Visser, Gerard. *In gesprek met Nietzsche*. Nijmegen: Vantilt, 2012.
Vuillot, Alain. *Heidegger et la Terre. L'Assise et le séjour*. Paris: Harmattan, 2001.

W

Waal, Frans de. *Good Natured. The Origins of Right and Wrong in Humans and Other Animals*. Cambridge MA: Harvard University Press, 1997.
Waal, Frans de. *Our Inner Ape. A Leading Primatologist Explains Why We Are Who We Are*. New York: Riverhead Books, 2006.
Waal, Frans de. *The Bonobo and the Atheist. In Search of Humanism among the Primates*. New York: W.W. Norton, 2013.
Wal, Koo van der. *Nieuwe vensters op de werkelijkheid. Contouren van een natuurfilosofie in ontwikkeling*. Zoetermeer: Klement/Pelckmans, 2011.
Welzer, Harald. *Klimatkriege. Wofür im 21. Jahrhundert getötet wird*. Frankfurt am Main: S. Fischer Verlag, 2008.
Wever, Patrick De. *Temps de la terre, temps de l'homme*. Paris: Albin Michel, 2012.
World Wildlife Fund. *Living Planet Reports* 2014.
Wotling, Patrick. *Nietzsche et le problème de la civilisation*. Paris: Presse Universitaire de France, 1995.
Wijssen, Frans and Sylvia Marcos (eds.). *Indigenous Voices in the Sustainability Discourse. Spirituality and the Struggle for a Better Quality of Life*. The Nijmegen Studies in Development and Cultural Change (NICCOS), no. 49. Münster: LIT Verlag, 2010.

Z

Zarka, Yves Charles and Caroline Guibert Lafaye (eds.). *Kant Cosmopolitique*. Paris: Éditions de l'Eclat, 2008.

Index

agriculture, impact on human beings   155 n.20
analogism   110
animals   5–9, 17–19, 68
   categories of   5
   and food   27
   functions, in Nietzsche's philosophy   6–7
   learning from   44–5
   terrasophy and   71, 76, 82, 88, 97, 98, 102, 108–11, 122, 134, 135
   Thus Spoke Zarathustra and   43, 47, 50, 52, 55–6, 60–1
Animate Earth (Harding)   84
animist culture   110
Anthropocene   126, 158 n.8
   Anthropos in   136–7
   and overman   137–8
   first phase of   126–7
   Gaia and   128
   'great health' and   138–9
   modernity and   129
   second phase of   127
   third phase   127
   views on   127–8
anthropocosmic worldview   119
Antichrist, The   16
Appadurai, Arjun   144
Appiah, Kwame   159 n.15
Argonauts   159 n.23
Astor, Dorian   152 nn.8–9, 153 n.6
Aurelius, Marcus   118
Autobiography, An (Gandhi)   100
autonomy   109, 117, 119, 128, 129, 151 n.6

Bataille, George   86, 156 n.31
Bestiaire de Friedrich Nietzsche (Brémondy)   5
Beyond Good and Evil   8, 16, 50, 111, 159 n.17
   on human beings and gardening metaphor   31
   on *kairos*   67
   on life skills   90
   on nationalism   37
   Shapiro on   63
Beyond Nature and Culture (Descola)   111
biocultural region   106, 155 n.26
bioregionalism   116
birds, importance of   6–7
blessed isle, myth of   48, 49
Blessed Unrest (Hawken)   90
body-object (*Körper*)   21, 22
Bolivia   115
Bonneuil, Christophe   158 n.8
Bonte, Sylviane   152 n.2
Brémondy, François   5
Brenner, Albert   10
buen vivir movement   115
Burkhardt, Jacob   28, 29

Campbell, Joseph   124
Case of Wagner, The   16
Chapelle, Gauthier   141
China   116
chronos   67
Cicerone, The (Burkhardt)   28
cities, significance of   143
Clifford, James   112–13, 115
climate and cosmos   34–8
   as interactive   38
   perspectives   36–7
   Stoics and   37
   temperate and tropical cultures   35
Colibri movement   156 n.38
collapsology   141
Colli, Giorgio   4, 22, 148–9 n.1
Commons' movement   155 n.25
compassion, significance of   58

compositionism 137
Confucius viii
contrast experiences viii
cosmopolitanism 158–9 n.15
cosmopolitan localism 119, 144–5
Crutzen, Paul 126
cultural endurance 112
cultures, ad multifarious and local 65–6

D'Iorio, Paolo 9–11, 151 n.4
Damasio, Alain 152 n.5
Darwin, Charles 40, 125
Dawn, The 4
Daybreak xi, 14, 15, 39
  'By circuitous paths' in 20
  on cosmology 38
  on gardening metaphor and human beings 30, 31
decolonization 112, 113, 115
Del Caro, Adrian 147 n.3, 148 n.6, 150 n.1, 159 n.17
Deleuze, Gilles 106, 147 n.4, 153 n.4, 156 n.31
Descola, Philippe 108–11, 131, 154 n.12, 159 n.16
Deussen, Paul 16
Dionysian Dithyrambs 16
Dohmen, Joep 32
dreams, significance of 52–4

eagle's cosmic view 74, 76, 143–4, 151 n.7
earth and human beings. See
  individual entries
eating, fasting, and diet 24–7
Ecce Homo 24–5, 97, 150 n.3
  climatic and cosmic perspective in 35
  experimental life in 85, 88
ecological awareness 72, 90, 97, 123
  local, in emerging economies 114–16
ecological consciousness xi, 91, 121, 122
ecological footprint 136
Écrits sur Nietzsche (Colli) 22
Ecuador 115
einverleiben metaphor. See
  incorporation process
Elkington, John 117
English landscape garden 28
Enlightenment mentality 118

Enlightenment morality 54
Epictetus 118
Epicurus 92–3, 152 nn.8–9
Escobar, Arturo 119, 144
ethno-linguistic groups 108
ethnological approach 106, 119–22
  becoming indigenous and 122–3
  of European history 111–13
  indigenous cultures and 113–14
  local culture and 107–11
  local ecological awareness in emerging economies and 114–16
  urbanization issue and 116–17
  West and local and 117–19
Evanoff, Richard 144, 155 n.26
experimental life 84–91
  children's attitude with 86
  consumer dynamic and 89
  dangers and difficulties impacting 86
  dissatisfaction with lifestyle and 85
  'no' and 'yes' skills with 88
  overman and 87
  self-discipline and 88

Finlay, Jocelyn 117
food 24
  German 25
  human influence and 26
  as link between humans and earth 25–6
  significance of 97–9
Förster, Elisabeth 149 n.13
Foucault, Michel 86, 156 n.31
freedom 28, 30–2, 75, 129, 141, 157 n.41
  experimental lifestyle and 6, 11–13, 15
  human dignity and 118
  individual 73
  social behavior and 75
  of thought 12
  West and local and 117–19
free spirit xiii, 10, 11, 14
French landscape garden 28
Fressoz, Jean-Baptiste 158 n.8
future, approach towards 140–2
futurity 67–8

Gaia cosmology 113, 126, 128,
    131–2, 152 n.3
  and Nietzsche's earth
    compared 132–4
Gandhi, Mahatma 100, 152 n.11
gardening metaphor 21, 22, 27
  cities and 143
  ethnology and 106–7
  and human beings
    attitudes and aptitudes 29–30
    combining contrasting
      forces in 31–2
    as organic and 30–1
  Komrij on 33–4
  landscape architecture and 28–9
  significance of 104–5
Gay Science, The xi, 4, 12, 13, 16, 39, 43
  on cities and urbanization 105
  on Epicurus 92–3
  on experimental life 84–5
  on food 24
  on 'great health' 139
  on landscape architecture 29
  on life 23, 38
  on morality and history 118
  on moral judgments 149 n.10
  on self transformation 94–5
  on Stoics 94
  on suffering 112
  on troubadours 42
  on yes-saying morality 99–100
genealogical analysis, of own past 96
  food and 97–9
  personal lifestyle and 100
Gerhardt, Volker 150 n.14
Germany 116
Gersdorff, Carl von 16
Gide, André 147 n.4
Glaeser, Edward 117
Global 200 107
Global Footprint Network 116
globalization xiii, 111–16, 119–
  22, 127, 144
Gosseries, Axel 153 n.14
'great events' 66
'great health', significance of 138–9
Guattari, Felix 147 n.4, 153 n.4, 156 n.31

Harding, Stephen 84
Hau'ofa, Epeli 114, 154 n.17

Hawken, Paul 90
Hegel, Friedrich 48, 102, 117, 118, 141
  and Nietzsche 64, 66
Heidegger, Martin 147 n.4
Human, All Too Human xi, 4, 17, 39, 81
  climate and cosmos in 35

incorporation process 22, 23, 27
India 115, 116
indigenous cultures 107–8, 121, 131
  becoming indigenous and 122–3
  importance and impact of 111–13
  local ecological awareness in emerging
    economies and 114–16
  perspective of 113–14
indigenous mentality 121
integrated culture 113
intergenerational justice 102, 153 n.14
Irigaray, Luce 147 n.4, 156 n.31
Italian landscape garden 28–9

Jonas, Hans xiv

Kafka, Franz 48
kairos 67
Kakarala, S. 157 n.41
Kant, Immanuel 54, 117, 118, 151 n.6
KARI-OCA 2 Declaration 154 n.14
Kaulbach, Friedrich 150 n.14
Kierkegaard, Søren 100, 133
Komrij, Gerrit 33–4
Körper. See body-object (*Körper*)
Köselitz, Heinrich 16, 19
Kuijpers, Andre 124, 125
Kultur/culture, notion of 153 n.6

land behind us metaphor 13
landscape architecture 28–9
language, significance of 83
Latin America 115
Latour, Bruno vii, xiii, 77, 127–9,
    131, 132, 137
Leib. See living/lived body
Le voyage de Nietzsche à Sorrente
    (D'Iorio) 9–10
life viii–xiv, 140–6, 156 n.31
  animals and nature and 3–5, 7–11,
    13–15, 17–20
  embracing 56
  experiences, in mountains 17

experimental 84–91
   natural world interaction and 21–5,
       27, 29, 30, 33–8
   significance of 23, 38
   terrasophy and 71–7, 80, 92–102,
       107–10, 113–14, 116–22, 124–6,
       130–4, 136
   in *Thus Spoke Zarathustra* 39–43, 45,
       49–52, 54–6, 58–62
living/lived body 21–2
Living Planet Reports 91
local–planetary paradigm 158–9 n.15
Lovell, James 124, 131, 152 n.3
Lovelock, James 126
lust to rule, views on 53, 54

MacLeish, Archibald 124
Makarov, Oleg 124
Mancini, Ezio 144, 145
Manschot, Henk 156 n.33
Manzini, Ezio 119
Margulis, Lynn 131
Marx, Karl 102, 118
Maurer, Reinhart 150 n.14
Meijer, Saskia van Goelst 152 n.11
mental food 26
Meyer, Lukas H. 153 n.14
Meysenbug, Malwida von 10, 11, 99
modern culture xii, 106, 107, 140
   criticism of ix, xi, 36, 72, 111
   as hyperactive 26
   individualism of 102
   naturalism and 109
   overman and 46, 102
modernity xii, xiii, 73–5, 111,
       129, 150 n.3
   circular time experience and 114
   earth and 73
   Gaia cosmology and 128
   human beings and 134
   humanism and 145
   man as abstract notion and 117
   social and cultural 87
   state and 120
Montaigne, Michel de 85
moral attitudes 159 n.18
moral consciousness 45, 100
morality 7–8, 90
   Enlightenment 54
   and history, linking of 118

   universal 75, 137
   yes-saying 99–100
mountains 15–20
   contact with animals in 17
   life experiences in 17
   mystical experiences with 18–19
   spiritual tradition and 17–18
Myths to Live By (Campbell) 124

Nachgelassene Fragmente 9, 14, 148 n.1
   on food and fasting 26, 27
   genealogical analysis of own
       past in 96
   on human nature and gardening
       metaphor 30, 32
   on mental food 26
   on overman 87, 150 n.3
   on Zarathustra 42
Nandy, Ashis 155 n.22
native wisdom, significance of 111
naturalism 109–10
Nelson, Jimmy 154 n.15
Niemeyer, Christian 150 n.3, 151 n.7
Nietzsche ix–xiii. *See also*
       individual entries
   and animals 5–9, 27, 76
   art of living well of 32–3, 81, 85, 101
   on body 30
   Dohmen on 32
   on free spirits 13
   Hegel and 64, 66
   journey to Sorrento 9–12, 48
   on modern culture ix, xi, xii, 26, 36,
       46, 72, 102, 106, 107, 111, 140
   on modernity xii, xiii, 73–5, 87, 114,
       117, 134, 150 n.3
   in mountains 15–20
   mystical experiences of 18–19, 40
   naturalization of 19
   on Plato 42
   by sea 13–15
   on silence of nature 14–15
   sufferings of 4
   Wotling on 35
Nietzsche's Earth (Shapiro) xii, 63
   gardening culture in 105–6
Nietzsche and Philosophy (Deleuze)
       106
*Nietzsche Briefwechsel. Kritische
       Gesamtausgabe* 12, 16, 17

Nietzsche et le problème de la civilisation (Wotling)   35
Nietzsche over de Menselijke Natuur (*Nietzsche on Human Nature*) (Dohmen)   32
Nussbaum, Martha   159 n.15

Ökologische Lebenskunst (Schmid)   125
On the Genealogy of Morals   7, 16, 50, 93
    emotions and alertness in   68–9
    ethnological approach in   106
    on future   142
    morality in   90
'On the Necessity of Gardening' (Komrij) (lecture)   33
Oranje, Wilfred   18
Our Common Future (Brundtland Commission) report   102
Overbeck, Franz   12, 16
overman   67–8, 98, 134–6, 148 n.6, 150 n.3
    Anthropos and   137–8
    bodily vision of life and   156 n.31
    bond with other generations and   102
    'brutal' body and   135
    ethnology and   108
    significance of   46–7, 54
    in *Thus Spoke Zarathustra*   56–62, 87
Overshoot Day   91

paradigm shift, significance of   40–1, 61
Par-delà nature et culture (Descola)   111
Pascal, Blaise   73
people, as multitudes   65
perception and expression   81–4
personal lifestyle, significance of   100
Picq, Pascal   108
pity, significance of   58
Plato   42, 118
power   17, 67–8, 106, 118, 137
    of attraction   89–90
    blessed isle myth and   48, 49
    expressive   42
    game of   47–51
    Gaia and   128
    of persuasion   110
    political   51
    shift of   120
    state and   48, 50, 115
    of transformation   137
    uncultivated natural   30
    of virtue   62
    will to   xi
primary naturalism   134

radical humanism, possibility of   145–6
Rahbi, Pierre   156 n.38
rationality   54, 65–7, 74, 93, 109, 118, 141
Ree, Paul   10
Revenge of Gaia, The (Lovelock)   126
Rio+20 conference   154 n.14
Rockström, Johan   127, 129, 158 n.9
Rohde, Erwin   12
rooted cosmopolitanism   159 n.15
Rousseau, Jean-Jacques   32
Royal, Ségolène   100

Safranski, Rüdiger   19
Schmid, Wilhelm   125
Schutter, Olivier de   155 n.21
Scritti su Nietzsche (Colli)   22
sea, concept of   13–15, 151 n.4
selfishness, views on   54
self-transformation   81, 91–6
    Epicurus and   92–3
    pair of scales metaphor and   92
    Stoics and   93–4
Séméria, Yves   152 n.2
Seneca   118
Servigne, Pablo   141
seven seals, significance of   55
sex, views on   53, 54
Shapiro, Gary   xii, 63–5, 69, 105–6, 147 n.3, 149 n.12, 153 n.4
    on futurity   67–8
    on Nietzsche
        alternative thoughts   65
        philosophy of earth   64
Shiva, Vandana   115, 155 n.22
silence, of nature   14–15
Sloterdijk, Peter   xiii, xiv, 77, 127, 128
snake's cosmic view   76, 151 n.7
sovereignty, significance of   128
Stafford, Tom   124
state   36, 64–6, 106, 137
    as end of history   66
    globalization and   115, 120
    fire hound as symbol of   48, 50
    Hegel and   64, 66
    as hypocrite hound   66

modern 114–15
power and 48, 50, 114–15
Stengers, Isabelle xiii, 77, 127–9, 131, 132
Stevens, Raphael 141
Stiegler, Bernard 89
Stoics 37, 93–4, 118, 152 n.8
Suransky, Carolina 153 n.14
sustainable development 153 n.14, 154 n.14, 155 n.20
sympathy, significance of 58

teleology and meaning, comparison of 102
Terra Lingua 107, 108
terrasophy 62, 69, 71, 74–8, 133. *See also* Anthropocene; ethnological approach
animals and 71, 76, 82, 88, 97, 98, 102, 108–11, 122, 134, 135
life and 71–7, 80, 92–102, 107–10, 113–14, 116–22, 124–6, 130–4, 136
Thoreau, Henry David 31
Thus Spoke Zarathustra xi, 79, 130, 134, 135, 140, 150 n.3
Act four (overman) 56–62, 87
Act one (on road with Zarathustra) 44–7
Act three
dreams 52–4
time and eternity 55–6
Act two (power game) 47–51
experimental lifestyle and 6, 8–9, 16, 19, 36, 39–40
mountain experiences in 18
Nietzsche's Earth and 63–9
orientation towards future in 101–3

as stage play 43
stylistic aspects of 41–2
as tragedy 42–3
time
circular experience of 114
and eternity 55–6
Tongeren, Paul van 17, 61, 150 n.13
totem culture 110
Triumph of the City (Glaeser) 117
truth pathos 83
Twilight of the Idols 16

Übermensch. *See* overman
Uganda 115
UNESCO 155 n.20
United Nations Development Programme (UNEP) 154 n.9
universal rights, in local contexts 157 n.41
urbanization issue 116–17

van der Merwe, J. C. 153 n.14
Van Gogh, Vincent 83, 152 n.2
vitality, notion of 25
voluntary beggar, significance of 58–60

Waal, Frans de 159 n.18
Wanderer and His Shadow, The 4
gardening culture in 105
Weiming, Tu 118–19
West and local 117–19
World Wildlife Fund (WWF) 107, 108
Wotling, Patrick 35, 153 n.6

yes-saying morality 99–100

Zeronauts, The (Elkington) 117

www.ingramcontent.com/pod-product-compliance
Lightning Source LLC
Chambersburg PA
CBHW070639300426
44111CB00013B/2172